BASIC GUIDE TO
Accident Investigation and Loss Control

WILEY BASIC GUIDE SERIES

BASIC GUIDE TO
Accident Investigation and Loss Control

Jeffrey W. Vincoli, CSP

JOHN WILEY & SONS, INC.
New York Chichester Weinheim Brisbane Singapore Toronto

Library of Congress Cataloging-in-Publication Data:

Vincoli, Jeffrey W.
    Basic guide to accident investigation and loss control / Jeffrey
W. Vincoli.
      p. cm.
    Includes bibliographical references (p.   ) and index.
    ISBN 0-471-28630-3
    1. Industrial accidents—Investigation. I. Title. II. Series.
HD7262.25.V56   1994
363.11'65—dc20
                                                     94-6092

10 9 8 7 6 5 4

In loving memory of Thomas Phillip Smeriglio, Sr. (1899–1976) whose great love for life, devotion to family, and unshakable faith in his God and country continue to inspire and motivate those who were fortunate enough to have known him.

Contents

Preface   **xi**

Acknowledgments   **xiii**

Part I   **The Accident Investigation Process**   **1**

INTRODUCTION TO PART I   1

Chapter 1   **Principles of Investigation and Loss Control**   **5**

INTRODUCTION   5
BACKGROUND—UNDERSTANDING
  ACCIDENT INVESTIGATION   6
ESTABLISHING INVESTIGATION
  PARAMETERS   9
SOURCES OF ACCIDENTS AND
  INCIDENTS   11
THE DOMINO EFFECT   15
SUMMARY   26

Chapter 2   **Responsibilities in the Investigation Process**   **28**

INTRODUCTION   28
THE MANAGER'S ROLE AS
  INVESTIGATOR   29

THE ROLE OF THE SAFETY
   PROFESSIONAL   32
ESSENTIAL ELEMENTS OF AN
   INVESTIGATION PROGRAM   33
SUMMARY   40

Chapter 3   **Investigation Planning and Preparation**   **42**

INTRODUCTION   42
THE PLANNING PROCESS   43
INVESTIGATOR SELECTION AND
   TRAINING   57
SUMMARY   63

Chapter 4   **Accident Response Actions**   **65**

INTRODUCTION   65
RESPONSE ACTIVITIES   66
PUBLIC RELATIONS   77
WITNESS INTERVIEWS AND
   STATEMENTS   83
THE ACCIDENT INVESTIGATION
   REPORT   99
DOCUMENTING LESSONS
   LEARNED   107
LEGAL ASPECTS OF ACCIDENT
   INVESTIGATION   108
SUMMARY   114

Chapter 5   **The Human Element**   **117**

INTRODUCTION   117
HUMAN ERROR VS. HUMAN FACTOR   118
PHYSICAL AND PSYCHOLOGICAL
   ASPECTS   120
THE SUPERVISOR'S ROLE   124
SUMMARY   126

Part II   **Tools and Techniques for Investigation**   **129**

INTRODUCTION TO PART II   129

Chapter 6   **The Accident Investigation Kit**   **131**

INTRODUCTION   131
INVESTIGATION KIT ITEM SELECTION   132
ADDITIONAL INFORMATION   138
SUMMARY   138

Chapter 7   **Photography in Accident Investigation**   **140**

INTRODUCTION   140
PLANNING ACTIVITIES   141
USES OF PHOTOGRAPHY   143
TYPES OF PHOTOGRAPHIC EQUIPMENT   144
THE INVESTIGATION KIT   148
METHODS AND TECHNIQUES   150
PHOTOGRAPHS AS EVIDENCE   159
THE PHOTOGRAPH LOG   160
SUMMARY   161

Chapter 8   **Collection and Examination of Records**   **164**

INTRODUCTION   164
SOURCES OF DOCUMENTED EVIDENCE   165
SUMMARY   170

Chapter 9   **Maps, Sketches, and Drawings**   **171**

INTRODUCTION   171
PLANNING REQUIREMENTS   172
POSITION-MAPPING   172
SUMMARY   181

Chapter 10   **System Safety Applications**   **182**

INTRODUCTION   182
FAULT TREE ANALYSIS   183
SUMMARY   195

Appendix A   **Sources of Additional Information/Training**   **197**

PROFESSIONAL ORGANIZATIONS   197
PUBLICATIONS—BOOKS   203

PUBLICATIONS—PERIODICALS   204
TRAINING SEMINARS/
    ORGANIZATIONS   206

Appendix B   Acronyms and Abbreviations   208

Appendix C   Accident Investigation Checklist   211

Appendix D   Case Studies in Accident Investigation   216

Glossary   224

Bibliography   232

Index   235

Preface

The Basic Guide to Accident Investigation and Loss Control is the third in a series of *Basic Guide* books that focus on topics of general interest to today's practicing safety and health professional. As with previous books in the series, this text has been designed to provide the reader with a fundamental understanding of the subject and, it is hoped, foster a desire for more advanced training and information. However, this particular *Basic Guide* differs from its predecessors in that it is not written solely for the practicing safety and health professional, but will serve additionally managers and supervisors who find themselves leading an accident investigation without knowledge of the fundamental principles of the investigation process.

While a great deal of information is presented in this text, it is not the objective of *Basic Guide to Accident Investigation and Loss Control* to provide or impart expertise in the field of accident investigation beyond that of novice. Those practitioners who require a more comprehensive and detailed instruction will not be satisfied with the information presented here, in a *basic* guidebook. However, those readers who are interested in acquiring a fundamental understanding of the accident investigation and loss control process should find this material useful. While there is nothing terribly new about the information presented here, it is hoped that the seasoned professional investigator will find this material enjoyable and, at the very least, refreshing.

In short, managers, supervisors, engineers, executives, property-casualty insurance personnel, quality inspectors, technicians, safety and health personnel, and trainers—all will obtain some benefit from the information contained in this text.

Acknowledgments

The author wishes to thank the following individuals and organizations who have assisted in the development of this publication:

George S. Brunner and Susie Adkins for their contributions and suggestions, and for proofreading this text.

Steven S. Phillips and Frank Beckage for helping to ensure the overall accuracy of the information and materials presented.

Tom Guilmet and the Central Florida Chapter of the National Safety Council for their contribution of reference materials.

Edwin P. Granberry, Jr. for his contribution of case study materials.

Alex Padro, Ken McCombs, Barbara Mathieu, and the staff at Van Nostrand Reinhold for their support, suggestions, and encouragement.

Kate Scully and the very competent professionals at Northeastern Graphic Services, Inc. for the composition and layout of this publication; for their advice, suggestions, and editing; and for making this Basic Guide Series the professional publication it is growing to be.

Finally, a special thanks to Jack M. Grass without whose assistance, suggestions, and contributions this text would not have been possible.

I

The Accident Investigation Process

INTRODUCTION TO PART I

Virtually all aspects of daily living involve a degree of *risk* that some event or action may not yield the result originally intended. An unintended result may be termed an *accident*, and when it has a negative or adverse impact on the overall task, plan, or event, it is often referred to as a *loss*. In everyday business operations, where profit margins are measured by the amount of return on original investment, *investigation* of such events becomes a critical component of *loss control*. Because an accident and, more specifically, any resulting loss (including injuries) will ultimately affect the profit base of a business, the fundamental purpose of accident investigation is to *identify cause and recommend corrective actions to preclude the possibility of future similar events*. When injury and human suffering result from an accident, identifying causes and correcting them becomes even more essential, so that no one else will fall victim to the same circumstances.

Typically, the person most often charged with the responsibility of accident investigation is the manager or line supervisor, and rightfully so. For it is the management function of any organization that is best situated, with both the authority and responsibility, to discover the true

1

cause of those loss events that have detracted from the profit margin of their enterprise. Since line managers are often held accountable for accidents that have occurred in their area of responsibility, they frequently become an integral part of the investigation process. However, guidance and instruction in proper accident-investigating techniques are seldom provided and, in most cases, the purpose of loss control is not well understood. For this reason, accident investigations often become nothing more than witch-hunts, seeking only to discover who was at fault so that blame may be assigned. Such investigations often fail in their attempt to meet the primary objective of a true accident investigation—*to prevent the occurrence of future similar events*—and are therefore nothing more than counterproductive exercises.

Part I of the *Basic Guide to Accident Investigation and Loss Control* , therefore, will focus on the investigation process, the nature of accident investigation and loss control, the importance of proper reporting, and the involvement of management in the investigation process. As an introduction, Figure I-1 shows those activities that occur as strictly *investigative* and those that are primarily *loss control.*

The early chapters of Part I will present the fundamental principles of investigation and loss control, the essential elements of investigation planning, investigator training, and proper response actions immediately following an accident event. Appropriate public relations during and following an accident occurrence, an important aspect of the overall investigation and loss control process, will also be addressed. Chapter 5 will conclude Part I with a brief discussion of the human element in the investigation process and the importance of ensuring proper consideration of the many factors that affect optimum human performance.

Since this text is concerned primarily with the investigation of accident events for *cause determination,* Part I will purposely avoid discussion on investigation of worker's compensation accidents. While the investigation of worker's compensation accidents might also involve loss control, the reasons for such investigations are very different. Investigation of worker's compensation accidents is typically focused on *claim verification* rather than prevention. Subsequently, worker's compensation investigations are primarily based upon an assumption of or speculation about deception and usually utilize controversial discovery methods, such as surveillance and unauthorized intrusions into an individual's private life in an attempt to verify fraudulent activities. Although these actions may be deemed necessary to substantiate a claim, worker's compensation investigations exceed the limits of this text. The reader is referred to the appendices for sources of additional information on worker's compensation and

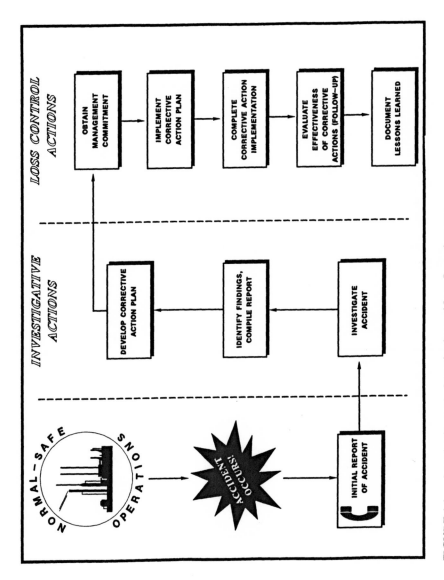

FIGURE 1-1. Introduction to the Investigation and Loss Control process.

3

other more advanced topics related to accident investigation and loss control.

Upon completion of Part I, the reader will have developed a fundamental understanding and basic appreciation of accident investigation and, more importantly, how it can be used to ensure subsequent losses are kept to their minimum.

1

Principles of Investigation and Loss Control

INTRODUCTION

Before attempting to discuss the process of accident investigation, it is important first to examine the fundamental principles associated with the analysis of such occurrences. Essential to the entire investigation process is the *proper reporting* of all accidents. This point cannot be overemphasized: The investigation of any accident event will never progress unless that event is first properly reported to the appropriate management element within an organization. Therefore, a *formal policy* requiring the proper and consistent reporting of all accidents is one of the most important principles of any accident investigation program. This simple precept to successful investigation is frequently overlooked, however, often because of the natural fear many people have of being associated with an accident event. It is understandable that the very idea of investigating a situation that has resulted in some degree of loss fosters apprehension among persons involved in or witness to an accident. It is all too common that investigations fail to identify the reasons an accident occurred because the investigator focused on assigning blame rather than determining cause. People are reluctant to report any event which might reflect unfavorably on their own performance or that of their department. In many cases, they simply do not wish to interrupt work production by reporting an accident. For these reasons, many accidents go unreported, are not investigated and, subsequently, are often destined to be repeated. Without complete reporting of accidents and losses, followed by a comprehensive investigation into the cause of such losses, management will have no exact knowledge of the extent and nature of the conditions which have downgraded the efficiency of their

many are afraid to come forward because they fear they will be blamed

5

business enterprise to the point where a loss occurred. It is therefore crucial to the investigation process that the investigator resist the natural tendency to focus on the finding of fault and assigning of blame. Human nature forces us to be somewhat suspicious and skeptical, traits not necessarily bad, but traits that should be used in a positive way during an investigation so that the accident analysis will be as beneficial and productive as possible.

BACKGROUND—UNDERSTANDING ACCIDENT INVESTIGATION

Until progressive safety concepts were inaugurated into the mainstream of business operations, virtually every safety rule was the result of some previous accident. This practice of locking the barn door *after* the horse escaped was an attempt to prevent the same accident from happening again. To a large degree, this after-the-fact method of safety and loss control was successful in the formulation of some of the early safety rules now standard practice in industry. But detailed investigation of the true cause of an accident event was still not a primary motivation during those early years of loss control. Managers usually promulgated safety rules *after* an accident, without any real investigation, in an attempt to prevent any further interruptions in the production process. At one time, a human fatality was generally required before any formal investigation would be initiated by management. This primitive managerial approach to accident investigation was frequently referred to as "*blood priority*" by the safety professionals of that era. Very simply, if there was no blood spilled, then there was no real priority (or budget) for any action, and even less management interest.

This lackluster approach to the prevention of accident recurrence is not a recent development. In fact, it can be traced as far back in history as the Babylonian Empire, in 2100 B.C., when the *Code of Hammurabi* prescribed indemnification for major injuries or death. Under the Code, if a person's negligence led to the injury or death of another, then the negligent party would loose a hand, or even a head, to match the loss of the injured. The Code's primary purpose was to assess blame and provide indemnification (or revenge), rather than to determine the cause of the accident itself.

Unfortunately, today, more than 4,000 years later, a similar though less dramatic approach to accident prevention still exists in some places. It is not uncommon to find that a business enterprise that has had no history of serious accidents, often has no established safety program either. Where investigations focus only on the *immediate causes* of each accident (rather than identifying the much deeper or *root causes*), and action plans stop at resolving not to do the same thing again, there has been little change in reducing either the frequency of accidents or their resulting severity.

Accidents, Incidents, and the Management of Risk

Up to this point, the term *accident* has been used frequently with the general understanding that most readers have a fundamental idea of the term's meaning. However, in order to ensure a common understanding of the term, the following definition is provided: As used in this text, an *accident* is an unplanned and therefore unwanted or undesired event which results in physical harm and/or property damage. The use of the word *unplanned* in this definition is meant to imply that no one truly intends to have an accident by purposely planning for its occurrence. However, it could be argued that organizations that purposely trade off safety margins for other more profit-oriented considerations might actually be planning for failure. This is more a matter of opinion than fact. While it is true that management may expect to incur accidental loss as part of the daily risks associated with business operations, it would not be fair to say that management *plans* for such events by totally ignoring sound safety policy. Rather, it would be more accurate to state that decisions made to *manage risk* require planning and that planning for failure is not normally the focus of risk management.

When accidents do occur, they are usually the result of one or more failures in a system or process, or the result of contact with a source of energy above the stress threshold of a particular body or structure. Situations where such contact is possible are often termed *hazardous*, and those conditions which permit the contact are considered *hazards*. Hazards are therefore dangerous conditions, either potential or inherent, which can interfere with the expected orderly progress of a given activity. Management often has some control over the amount of exposure to hazard risk via the implementation of policies and practices that can effectively alter conditions like the area of energy absorption or the duration of contact. Management can also control the amount of hazard risk associated with a given task or function. The value of such an energy control concept cannot be overemphasized because management can utilize it as an effective tool in the early recognition and reduction of hazard risk. For example, a grinding machine operates with a high level of energy and, therefore, a high potential for injury or death should that energy be transferred to another body through unplanned (accidental) contact with the wheel during normal operations. Properly guarded, the wheel still has the same level of energy (i.e., the same level of hazard), but the potential to transfer that energy under normal operating conditions is substantially reduced. Hence, through the placement of a physical control such as a guard, together with an enforced management policy for guarding machinery, the hazardous condition or risk of exposure to the hazard can be virtually eliminated.

Unplanned events that could have but did not result in damage or injury are still considered as undesired or unwanted occurrences since they interrupt the normal, expected flow of a task—these are referred to as *incidents.* Since these events can present a significant level of risk while not meeting the full criteria for accident (having resulted in no detectable injury or damage), they are also often referred to as *near accidents,* or, more commonly, *"near misses."* It is the manager who must consider whether such events constitute losses or significant problems for the business enterprise. For instance, if production is disrupted for several hours or even days because a typing error resulted in the acquisition of a wrong part for a machine, and installation of that part produced improper operation (even before any material was ruined or injuries occurred), then this type of incident would represent a real and significant loss. Undesired events that do not result in physical damage or injury can still downgrade the efficiency of an activity. These events, termed *incidents* in safety and loss control, should therefore be investigated with the same resolve as are accidents. More specifically, an ultimate goal of any investigation program should be to examine all incidents for their accident loss potential and investigate those which are considered to have significant risk levels with respect to a company's goals and policies.

Few people would question the premise that sane persons do not desire to injure themselves or lose the use and value of property through damage. But it is also true that businesses today are built upon economic and engineering risks, and success is built on management of that risk. This risk environment can foster a competitive chance-taking attitude. However, the costs of injury and damage losses, changing social implications of worker deaths and illnesses, and modern legal liability decisions make accident risk-taking an increasingly undesirable practice.

In order to reach any goal or objective, businesses today need to develop an increased sensitivity to daily risk-taking activities, their managers reflecting this sensitivity on behalf of their people in decisions and actions. Hence, *risk management* is as important a factor in business success as any other management function, and the investigation of accidents and incidents becomes essential in the determination of risk levels which exist within the operation of that business enterprise. Since risk is often evaluated in terms of its potential cost to business operations, a major concept of risk management is the consideration of injury, illness, and property damage as significant *costs* of accident events. After all, sources of physical harm to personnel are also sources of compensation, liability, and tort claims. Likewise, sources of property damage/loss are also sources of lost production, equipment replacement, and the loss of a competitive edge. Herein lies the essence of loss control: The proper investigation of accident cause with appropriate follow-up and adequate management involvement, all contribute to the control of lost resources (people, equipment, etc.) and the subsequent loss of business opportunities. Also, the concept of risk management is as important to the accident investigation process as it is to the control of losses because, as mentioned earlier in this chapter, risk-taking involves willful activity (or the willful lack of activity) on some manager's part. In order to evaluate risk properly in a given situation, management first requires a definable probability of an accident and a definable potential for losses. In the absence of such information, management decisions become increasingly subjective, which in and of itself is somewhat risky. It is the degree of subjectivity when calculating these probabilities that leads to many accident events. These *calculated risks* often are not well calculated at all but are more frequently *educated guesses*. It is the accident investigation that must determine the value of each risk-taking action and the exact effect these actions might have on the final outcome or loss event.

Simply stated, an *accident investigation* is a methodical effort to collect and interpret facts. It is a systematic look at the nature and extent of the accident, the risks taken, and loss(es) involved. It is an inquiry as to how and why the accident event occurred. Since a major function of the investigation is to consider what actions can be taken to prevent similar events from occurring in the future, it is also a *planning process* to explore actions that should be taken to prevent or minimize recurrence of the accident. Chapter 3 will explore the planning process in greater detail to demonstrate its value in the reduction of loss potential.

ESTABLISHING INVESTIGATION PARAMETERS

For an accident investigation to be as productive a tool as possible, management must first establish appropriate *parameters* within which the investigation is to be conducted. The establishment of such parameters for investigation is possibly one of the most important executive decisions in any business enterprise. Investigation and subsequent feedback are elements of management control. Unless management establishes these controls, they will be blind to the effects of accident loss on their business. Much of what is known today about accident prevention and loss control was, in fact, learned from loss incidents which were properly reported and investigated. However, it is probably safe to say that an even greater body of knowledge has been lost over the years simply because no one bothered to ask the question "Why?"

Effective parameters for accident investigation may include such aspects as the types of occurrences that will require reporting and investigating, what elements of a business operation or function will be investigated and to what extent, how accidents and incidents shall be formally reported, and what use shall be made of the information reported. Factors which will affect the selection of investigation parameters are varied and diverse, depending upon the perspective of the interested party, and may include any of the following examples:

1. *Stockholders and Owners* have understandable proprietary interests in liability for the safety of the workplace and in the value of real property.
2. *Executives and Managers* are interested in sales, work processes, productivity, general liability, and the direction and control of the business organization.
3. *Suppliers* have an interest in product liability with regard to machines and materials they have provided.
4. *Insurers* have an interest in obligations to pay benefits, damage repairs, and work-interruption settlements.
5. *Supervisors* are interested in the management of the work tasks associated with day-to-day operations.
6. *Workers* have a personal interest in their own safety, health, and welfare.
7. *Unions* are interested in a worker's rights, job security, compensation and benefits, and the general welfare of the workforce.
8. *Communities* are interested in hazards to residents and the economic impact of a business enterprise on the local economy.
9. *Government Agencies* are interested in compliance with regulations, codes and standards, and any impact on national welfare.

Regardless of the particular interests of the involved parties, without a definitive and credible reporting system with clearly established parameters, accident reporting will follow the subordinate manager's perception of the requirements. For example, in a work environment where self-reporting brings punishment or negative results, the obvious losses will typically be the only ones reported because they cannot be hidden. Minor losses, which could be the basis for loss control, will often go unmentioned. In fact, in such an environment, investigations of major accidents frequently reveal the occurrence of earlier incidents that had been dismissed as insignificant, rare occurrences, acts of God, or as of little consequence. However, properly investigated and evaluated for potential loss, such incidents could have been the basis for management actions to prevent later, more serious losses. Ironically, the repercussions which the manager feared and attempted to subvert by not reporting an incident, often come about anyway as a direct result of such actions to avoid minor censure. Another paradox often occurs in many organizations, large and small, in that each level of management seeks full reporting from subordinate managers or workers, but also seeks relief for themselves from any and all reporting to a higher level.

The fact that an accident occurs is a strong indication that a manager somewhere within the organization has made a bad decision. A personnel manager may have erred in selection and training, a purchasing manager may have misinterpreted purchase specifications, a work center manager may have erred in job assignment, or an executive may have erred in the allocation of sufficient resources for the task. At any level, *erroneous decisions* can and often do lead to an accident or loss situation. The nature of accident investigation requires an analysis of all possible reasons such decisions have been made. Faulty communications, lack of adequate information, and improper training are examples of causal factors which are often repeated during the investigation of accidents and incidents. Repeated causal factors are characteristic of high-level management failure to control losses through the establishment of effective investigation parameters. Disciplined, methodical reporting can be fruitful, productive, and beneficial to an entire organization. However, a misguided, haphazard investigation of those reported accidents will only increase the potential for loss. Therefore, a credible investigation policy with clearly established parameters is an important and useful tool for managers and workers alike.

SOURCES OF ACCIDENTS AND INCIDENTS

In addition to understanding the fundamental principles of accident investigation and the nature of accidents, incidents, and the management of risk, the accident investigator must also understand the *sources* of these undesired events. Without such understanding, it will be difficult to relate the methodical tracing of evidence to the basic accident causes, which is essential to the overall investigation process. Hence, before any presentation and discussion of the methods and techniques used for investigation can be attempted, the sources of downgrading incidents which lead to accidents and loss shall first be explored.

Sources of Potential Loss

In accident investigation, it is commonly understood that a combination of factors or causes must usually come together under just the right circumstances to bring about these undesired events. Seldom, if ever, is there a single cause of a loss incident that affects such diverse operational aspects as safety, production, quality, management, etc., all at the same point in time.

It is not logical even to assume that a single causal factor could have drastic effects across an entire organizational structure. All factors which could have worked together to downgrade the efficiency of a particular business activity must, therefore, be properly investigated and recommendations provided, to ensure that such events do not occur again.

As complicated as the problem of accident cause and prevention may appear, progress has been made over the years, proving that it is quite possible to prevent or control the causes of downgrading incidents. The aviation and aerospace industries are perfect examples. If many thousands of potential hazards had not been eliminated or controlled adequately, man would have never made it to the moon and back safely. Likewise, if hazard recognition and control were not practiced daily in the airline industry, then flight would not be considered the safest mode of transportation, as it is today. While the enormous amount of resources invested in both the aerospace and aviation industries may not be available to everyone, there is well-documented evidence that high levels of accident risk reduction can be achieved by the average business through proper investigation and analysis of accident cause.

Information on accident causation and reduction, developed after many years of review and analysis, has led management leaders to conclude: (1) Incidents which downgrade business performance are in fact caused—they do not *just happen*; and (2) The causes of downgrading incidents can be determined and therefore controlled. To understand the circumstances which lead to the causes of these undesired incidents, management must first examine the five principle elements of their organization that are considered the sources of such events. Figure 1-1 shows the five elements of people, equipment, material, procedures, and the work environment which must all interact together for successful business operations. However, when something unplanned and undesired occurs within either of these elements, there is usually some adverse effect on any one or all of the other elements that, if allowed to continue uncorrected, could lead to an incident or accident and subsequent loss. It should follow, therefore, that all five elements must relate properly to each other in order for a business to be successful.

To further understand how these five elements can become sources of most, if not all, downgrading incidents in business operations, each shall be briefly examined.

People

Obviously, the people of any successful organization are its greatest resource. On both the employee and management levels, a business is only as good as the combined talents and abilities of its people. Unfortunately, statistics show that a high percentage of incident causes have been attrib-

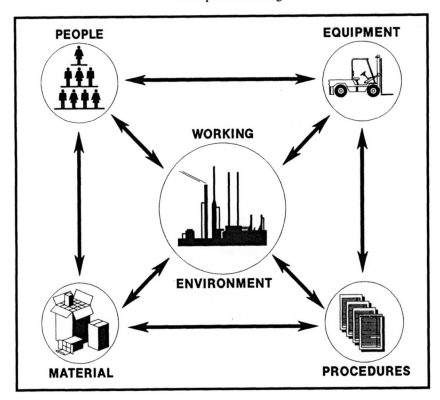

FIGURE 1-1.   The five elements that must interact for successful business operations.

uted to the human element. It must be clear, however, that proper accident investigation does not end with a simple causal determination of *human error*. Influencing factors that can affect human behavior and performance must also be examined and evaluated for cause. These may include factors such as education and training, or motivation, or tools provided to perform assigned duties. Employees are the human element directly involved in most accidents, since what they do or fail to do is often viewed as the immediate cause. However, in the daily operation of a business enterprise, it is the employee-management relationship that greatly influences the actions of individual employees. If, for example, inadequate training is provided by management to an employee prior to the performance of a new task and the employee then makes an error resulting in an accident, it would not be a complete investigation to conclude that the employee error was the only cause of the accident. A proper investigation would reveal that management erred by providing faulty training to the employee, which was the primary cause of the accident event. The relationship between human error and human factor is an extremely important concept that will be discussed at great length in Chapter 5.

Equipment

Equipment includes the tools and machinery employees must work with in order to accomplish their assigned duties. Machinery could involve such items as drill presses, lathes, and grinders, as well as cranes, forklifts, and automobiles. This element or subsystem of business operations has been a major source of incident causes since the early 1900s and has resulted in the promulgation of hundreds of laws requiring mechanical safeguarding and operator training.

In more recent years, the improper design of controls and displays on complex power machinery and equipment has been cited frequently as the primary source or cause of many downgrading incidents. The fact that a person's error in interpreting a display led to the incident should not detract from the primary causal factor of faulty equipment design. It should be noted that the above emphasis on power equipment is not meant to belittle the source of accident causes provided by such simple tools as wrenches, hammers and chisels. The improper use of hand tools, calibration instruments, gauges, or even a ladder to accomplish a given task can also adversely affect the outcome of the job to the point of incident or accident. Using a screwdriver where a chisel is required, or the handle of a wrench when a hammer should be used, is not only an indication of improper employee awareness, but also it may reflect a less-than-adequate management policy, improper tools available for a task, or misguided employee motivation. These examples demonstrate how adverse conditions in the first element, *people,* can and often will interact with the second element, *equipment,* (or vice versa) to result in a downgrading incident.

Material

The material people use, work with, or make provides another major source of incident causes. Nearly all of the thousands of product-liability cases that end up in American courts each year allegedly claim this element as the source of serious injury for a staggering number of people. Material can be sharp, heavy, hot, cold, toxic, or defective. In all cases, the material element of the business system can be a major source of energy contacts that result in downgrading incidents. Here again, since people must interface regularly with materials in order to perform their tasks, an examination of the relationship between potential incident sources becomes extremely important in the incident investigation process. When one considers that *people* often use *equipment* to process *materials* in daily operations, the complexity of incident source relationships becomes even more evident.

Procedures

In most organizations, policies and procedures are formulated and implemented with the intention of placing administrative limits or controls on employee behavior, on specific operating practices, and/or on defining acceptable company protocol. In many cases, exact procedures or operating instructions may be required for a complex machine or an intricate work task. If these procedures are less than adequate for the task or function to be performed, the employee may err in operating the equipment or performing the task steps, which could result in an incident. A faulty operating procedure is an indication of poor management involvement and control. Procedures must therefore be examined during an investigation to establish adequacy and determine their causal effect, if any, on the incident under investigation. Remember once again the other elements of accident source discussed thus far: *People* follow *procedures* to operate *equipment* and process *materials* to accomplish their assigned tasks.

Work Environment

The fifth potential element for incident source is the physical surroundings in which work must be performed. These include the buildings that house the people and the air they breathe. Environment is also associated with such items as lighting, noise levels, and atmospheric conditions (temperature, air quality, humidity, etc.). Since the previous four source elements (people, equipment, materials, and procedures) must exist within the work environment, adverse conditions in this fifth element could obviously affect any or all of the other four. The work environment represents the source of causes of an ever-increasing number of diseases and health-related conditions. These conditions affect people, which subsequently affects productivity. In fact, inadequate work environment is often cited as a major source of incident causes associated with absenteeism and poor work quality.

These five elements of the business operation (people, equipment, material, procedures, and work environment), either individually or in combination, produce the source of causes that contribute to a downgrading incident. Therefore, in the investigation of every accident or incident, management must ensure the proper consideration of the potential for involvement of any or all of them.

THE DOMINO EFFECT

It was established in the preceding pages that downgrading events which occur in one source element can affect conditions in one or more of the others, which, in turn, might lead ultimately to an incident or accident. This *cause and effect* relationship has often been described as the *domino effect*.

Figure 1-2 shows how the specific *phases* of the accident process, each represented by a domino, can lead to a loss. When each phase experiences a problem of significant adversity, it causes an effect on the next phase. As each phase affects the next, like dominos falling against one another, the ultimate result could be a loss. As shown in the illustration, the phases involved in the accident process include *management's* control of the organization, the *origin(s)* of the downgrading problems, the *symptoms* of those problems, the subsequent *contact,* and the ultimate *loss.*

To fully understand the domino effect, each phase (domino) in the sequence shall be examined separately.

Management—Lack of Control
The first domino in the sequence of events that could lead to a downgrading incident and subsequent loss is the lack of sufficient control by management. The word *control,* as used here, refers to one of the four main functions of any professional manager: planning, organizing, leading, and controlling. All of these functions pertain to the work that any manager performs on a daily basis. Whether the manager is a frontline supervisor or the president of the organization, these basic functions still apply. It does not matter whether work is directed at safety, quality, production, or cost. In all cases, the manager must plan, organize, lead, and control to ensure work is performed as required.

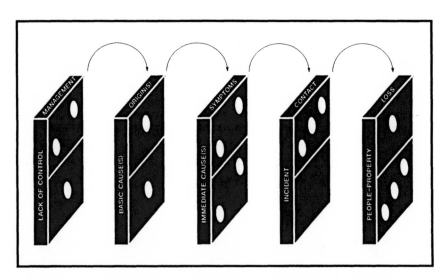

FIGURE 1-2.   The domino effect representing the *phases* of the accident process.

These four functions also relate to a loss control program. Supervisors can prevent a loss from occurring through planning and organizing such tasks as inspections, group meetings, and proper orientation for new employees. As leaders, supervisors can conduct investigations of problem areas or conditions and implement controls to prevent any further downgrading of operational efficiency. Supervisors can also exercise control by issuing rules and procedures for proper job/task performance, conducting analyses of specific tasks to identify the hazards and properly instructing those who will be required to perform the job. All these tasks are elements of management's involvement in prevention of downgrading events. The simple fact that many supervisors are unaware of the total work expected of them in the area of loss control, prevents them from doing their job properly. This alone can cause the first domino to fall, starting the sequence of events that will lead to loss.

By properly planning, organizing, leading, and controlling, a proactive supervisor can take the appropriate corrective actions *before* a potential loss occurs, rather than reacting *after* the loss event. Note that the frontline supervisor is the organization's primary point of control for losses that can occur during daily operations. However, if upper management does not practice the very same management functions and does not support the efforts of the frontline supervisor, then the domino will fall regardless of the best efforts of supervision. In short, supervisors who manage professionally:

- Know the standards of their loss control program;
- Plan and organize work to meet those standards;
- Lead personnel to desire to achieve those standards;
- Measure their own performance and that of their people;
- Evaluate performance in terms of loss control standards;
- Correct their own performance and that of their people.

A failure in any one of these areas could indicate a lack of management control, which could cause that first domino to fall and start the cause-and-effect sequence of events leading to a loss. For example, lack of management control begins when supervision has embraced an inadequate loss control program; or, even if the program is adequate, the supervisor may have inadequate knowledge of the program. Additional shortcomings develop when the program has established inadequate standards or the employee team fails to perform to adequate standards.

Origins—Basic Cause(s)
A lack of management control permits the existence of certain basic causes of incidents that downgrade the business operation. These causes have also been referred to as root causes, indirect causes, underlying causes, or real

causes, since the immediate causes (the third or next domino) most closely associated with the incident originate directly from these basic causes.

Basic causes are frequently classified into two major groups:

1. Personnel Factors
   Lack of knowledge or skill
   Improper motivation
   Physical or mental problems
   Personal problems (marital, financial, etc.)
2. Job Factors
   Inadequate work
   Inadequate design or maintenance
   Inadequate purchasing standards
   Normal wear and tear
   Abnormal usage

Those basic causes referred to as personnel factors explain why some people engage in substandard practices (the third domino) which lead to a loss. For example, a person may not follow a proper procedure if he or she has never been instructed to do so. An employee may not have the skill necessary to operate special equipment such as a crane safely and efficiently if proper training were not provided. Likewise, poor quality of work may result from the placement of a color-blind person on a job where excellent color vision is mandatory.

Similarly, the basic causes referred to as job factors explain why substandard conditions are created or may already exist. Equipment and material will be purchased and structures designed without proper consideration for loss control; if adequate standards do not exist or compliance with standards is not managed. Lack of proper maintenance of machines and equipment will result in substandard performance and unsafe conditions. Abuse and reuse of material, machines, and equipment can cause many conditions that result in waste and inefficient operation, thus presenting hazards to people and property.

Basic causes are clearly the *origin* of substandard acts and conditions. Failure to properly identify these origins of loss at this early stage in the sequence permits the domino to fall, initiating the possibility of a further chain reaction leading to the ultimate undesired occurrence, the loss.

Immediate Cause(s)—Symptoms
When the basic causes of downgrading incidents exist, they provide opportunity for the occurrence of substandard practices and conditions (often referred to as *errors*) that could subsequently cause the third domino to fall, leading directly to a loss event. Very simply, substandard practices or

conditions are any deviation from an accepted standard or practice. This could involve both acts by people and conditions related to physical objects. They usually occur as a result of some deeper causal factor (the second domino) associated with a lack of management control (the first domino) over a given situation or circumstance.

In the practice of safety and loss control, the immediate causes that permit the third domino to fall are often referred to as *unsafe acts* and *unsafe conditions*. Specifically, an unsafe act can be described as a violation of an accepted safe procedure (i.e., the standard) which could permit the occurrence of an accident. Likewise, the unsafe condition is an actual hazardous physical condition or circumstance which could directly permit the occurrence of an accident. Examples of some of the most common unsafe acts and conditions, leading typically to accident occurrences, are shown in Table 1-1. Whether or not these deviations are called substandard or unsafe, each is only a *symptom* of a basic or root cause that permitted such deviations to exist in the first place. The successful accident investigator does not stop at the determination of an unsafe act or condition as the immediate cause. The investigator must go deeper to understand what allowed such conditions to exist or what influencing factors led to the commission of unsafe acts. Only by doing so can the investigator determine exactly where the management process failed so that corrective actions can be implemented to preclude the

TABLE 1-1   Examples of common unsafe acts and conditions. (*Source:* American National Standards Institute Z16.2-1962, R1969.)

UNSAFE ACTS	UNSAFE CONDITIONS
OPERATING WITHOUT AUTHORITY	INADEQUATE GUARDS OR PROTECTION
FAILURE TO WARN OR SECURE	DEFECTIVE TOOLS, EQUIPMENT, SUBSTANCES
OPERATING AT IMPROPER SPEED	CONGESTED WORK AREAS
MAKING SAFETY DEVICES INOPERABLE	INADEQUATE WARNING SYSTEM
USING DEFECTIVE EQUIPMENT	FIRE OR EXPLOSION HAZARDS
USING EQUIPMENT IMPROPERLY	SUBSTANDARD HOUSEKEEPING
FAILURE TO USE PROTECTIVE EQUIPMENT	HAZARDOUS ATMOSPHERIC CONDITIONS
IMPROPER LOADING OR PLACEMENT	EXCESSIVE NOISE
IMPROPER LIFTING	RADIATION EXPOSURES
TAKING IMPROPER POSITION	INADEQUATE ILLUMINATION OR VENTILATION
SERVICING EQUIPMENT IN MOTION	
HORSEPLAY	
USE OF ALCOHOL OR DRUGS	

possibility of future similar events. When managers fail to discover the basic causes behind the symptoms, they will also fail to keep the third domino from falling, and the direct potential will then exist for an incident/accident and possible loss.

Incident—Contact

When unsafe acts and conditions are allowed to continue uncorrected, the opportunity for the occurrence of an incident that may or may not result in a loss also exists. Regardless of its outcome, the incident will be *undesired* since it was not intended and the final results of its occurrence are difficult to predict and are frequently a matter of chance. As mentioned previously, incidents that result in physical harm or property damage are referred to as accidents and usually involve a contact with a source of energy. However, it is important to remember that each incident, whether or not it results in loss, provides an opportunity to obtain information which could prevent or at least control a similar incident in the future.

For investigative as well as recordkeeping purposes, accidents are frequently categorized, according to the type of contact involved (ANSI provides some examples of these categories in its *Method of Recording Basic Facts*), as follows:

1. Struck Against
2. Struck By
3. Fall to Level Below
4. Fall on Same Level
5. Caught On
6. Caught Between
7. Caught In
8. Contact With:
   a. Electricity
   b. Heat
   c. Cold
   d. Radiation
   e. Caustics
   f. Noise
   g. Toxic or Noxious Substances
9. Overexertion (Overload)

When information obtained from the investigation of previous accidents is not recorded, categorized, and evaluated, the chance or likelihood of future losses cannot be prevented and the fourth domino will inevitably fall.

People/Property—Loss
Once the entire sequence of events has taken place and there is a loss involved with people and/or property, the results are usually chance events in that they often cannot be predicted with 100% certainty. This element of chance is involved in quality and production losses as well as those involved with losses which affect personnel safety, health, and security. Regardless of the particular business activity in which a loss may have occurred, losses can be considered minor, serious, major, or catastrophic. Depending upon the particular business activity, the determining factors for rating the *severity* of an accident loss can be the degree of physical harm and/or property damage and any resulting humane or economic aspects. Figure 1-3 provides examples of the types of losses which may occur in these various categories as a result of an accident. The severity of a particular loss event, in terms of minor, serious, major, or catastrophic, might be determined based on the degree of loss in any of the categories shown in Figure 1-3. When evaluating either the humane or the economic effects, investigators should be particularly aware that those factors readily apparent are usually indicative of much more serious and far-reaching aspects that are not so obvious. Much like the tip of an iceberg, the extent and size of the problems associated with accidents and losses are not easily seen or determined on the surface, but they are, nevertheless, there.

The Three Stages of Loss Control

Up to this point, each sequence (or domino) in the accident/loss process has been described and discussed separately. However, if the five sequences are grouped into three *stages of control* (Figure 1-4) entitled *pre-contact, contact,* and *post-contact,* it becomes even more obvious that the greatest opportunity to practice loss control occurs during the pre-contact stage. However, an evaluation of the remaining two stages will demonstrate that a significant opportunity also exists in each for management to implement loss control strategies to reduce accident risk potential.

Pre-Contact Stage
Loss control efforts at this stage or level include all actions performed to prevent the causes, as well as to detect and correct them, before contacts or incidents occur. It might also be referred to as the pre-incident stage of loss control. At this stage, management should be concerned with issues such as engineering practices in the design of new or modified facilities, purchasing practices in the procurement of new materials and equipment, and the development of work standards that could have loss potential implications. While frontline supervisors may not always have a specific responsibility to

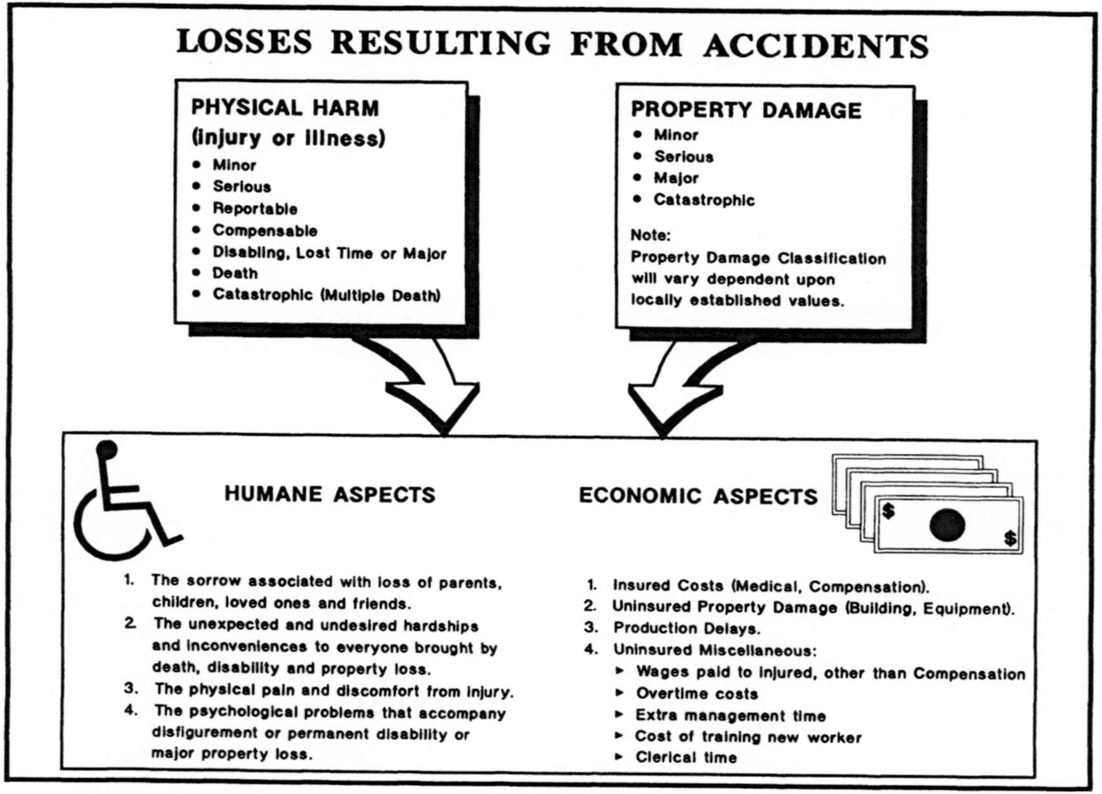

FIGURE 1-3. Examples of the types of losses that may occur as a result of an accident.

22

ensure these issues are addressed, their opinions and advice can be a valuable tool to those who have direct charge of such decisions. Clear lines of communication, therefore, are very important between supervision and other groups, organizations, and staff personnel. In fact, the supervisor's application of personal motivation practices, job indoctrination, standard job procedures, proper instruction, and the use and enforcement of rules in the work areas are all elements in the pre-contact stage that may actually prevent the first domino from falling.

Contact Stage

This is the point where the incident occurs that may or may not result in a loss, depending on the amount of energy transferred. It is important that management require the reporting of all non-loss incidents or "near misses" in order to implement appropriate corrective action before repetitive incidents occur that could have loss results. Proper management practices during the contact stage are also essential for the significant reduction of loss potential. For instance, a manager who ensures compliance with all protective equipment standards will reduce the potential effects of any energy contacts and hopefully prevent a loss from occurring. The astute supervisor can also make recommendations at the contact stage of control in an effort to minimize the amount of energy transferred. Some examples of such recommended measures that can be implemented to minimize losses

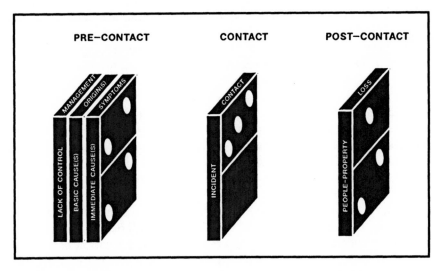

FIGURE 1-4.   The five sequences or phases of an accident grouped into the three *stages of control.*

involved with injury, illness, or property damage at the contact stage are as follows:

Hazard Elimination.   A supervisor can recommend alternate equipment or materials to eliminate any harmful energy source created by existing equipment or materials. For example, an electric motor might be used in place of rotating shafts and belts normally associated with power machinery, or a solvent with a higher exposure threshold or flash point can be used in place of more hazardous commodities.

Hazard Reduction.   Recommendations could be made to reduce the amount of energy used or released during the contact stage which will significantly reduce the risk of exposure to hazardous conditions. Water temperatures can be reduced to below-scalding levels in shower rooms, in-plant vehicle speeds can be controlled by the placement of speed bumps in passageways, etc.

Separation.   Whenever hazardous conditions are continuously present, a supervisor can separate persons or property from these conditions using both time and space to control potential exposure to such conditions. Hazardous materials can be stored in a locked or secured area away from personnel. The hazard is still there, but the risk of exposure can be controlled. Expensive equipment can be located off the floor level and out of traffic by placing it in a secure overhead position. The harmful effects of prolonged exposure to particularly noisy equipment can be controlled by providing adequate separation of personnel workstations located in the area or limiting the amount of time people are exposed to high noise levels each day.

Positive Controls.   When situations will not permit the elimination, reduction, or separation of hazards from personnel or property, supervision can still control the harmful effects of energy contact through the proper use of barriers or barricades between the source of energy and the people or property potentially exposed. Personal protective equipment (PPE), a guardrail on a loading platform, protective cement columns located around expensive stationary equipment, and insulation on noise-emitting machinery are examples of some of the types of barriers typically utilized to positively control hazard exposures.

Modification.   The areas of probable contact with hazardous materials or surfaces can be modified to control the potentially harmful effect of any such contacts. Foam padding can be installed on low ceiling points in stairwells or around sharp corners of countertops to minimize the risk of head injury or body bruises should contact occur. Modifications differ from barriers in that a barrier simply *protects* from contact with a hazardous condition that still exists, whereas a modification actually *changes* the condition so that it is protected in such a way that any contact would be less hazardous.

The recommended examples briefly discussed above are intended to reduce the loss potential *once a contact occurs,* but such actions will not prevent the contact events from occurring. It should be well understood that the prevention or elimination of exposure at the pre-contact stage should always be a primary management objective. In the real world of business operations, however, pre-contact elimination or prevention is not always entirely possible, and the use of loss control measures in the contact stage will provide further assurance that losses will be reduced. Such measures must always be complementary to, and not substitutions for, any pre-contact prevention efforts.

Post-Contact Stage
It is safe to assume that no member of the management team desires harmful or damaging contacts or incidents that result in a loss. However, although such occurrences can be reduced to a very low frequency level during the pre-contact and contact stages, an imperfect system or organization (and most are not perfect) will still have the potential for some type and degree of loss. Through thoughtful planning and organizing, management must be prepared to effect as much control as possible over the degree of loss during the post-contact stage. Even though it may be after-the-fact, there are still meaningful actions management can implement during this stage that might make the difference between injury and death or damage and total loss. These actions include but are not limited to the following examples:

Emergency Response.   The availability and immediate provision of emergency care has proven time and again to be an effective measure to reduce the frequency of death and disability in the business environment. There is no way to calculate the number of lives that can be saved or permanent disabilities averted each year if prompt, proper care were more readily available. Personnel should be trained and ready to cope with all types of potential emergencies that could arise in their work environments. They should know how to properly respond to a given situation, while preventing injuries to themselves and others, and minimizing property damage and loss. Such training typically includes knowledge and experience in evacuation and escape procedures, the proper use of fire-fighting equipment and protective devices, the sounding of alarms, and the emergency shut-off of specific equipment. Clearly, adequate loss control efforts during the post-contact stage could mean the difference between a minor occurrence and a major catastrophe.

Rehabilitation.   Injured workers can remain off the job for quite extended periods of time. All the while, the loss incurred by the accident is felt on the individual level, through human suffering and the loss of full earning potential, and on the total organization through lost production time, retraining and rework to compensate for the absent employee, and the payment of benefits while the employee loses time from work. A management policy that focuses on rehabilitation of the injured worker into other jobs or tasks will retain the employee's intrinsic value to the organization while reducing the overall loss resulting from an accident.

Prompt Repair Actions.   There are many occasions where prompt maintenance and repair of equipment can reduce the potential for greater loss or damage to property. Management should be alert for any evidence of property damage or premature wear-out of equipment. Small cracks in machines, equipment, or building structures can frequently be repaired easily in the early stages of their development.

Salvage Control.   Each year in the business world, millions of dollars worth of salvageable items are needlessly discarded because workers feel that these items are no longer useful. In reality, damaged equipment can often be repaired and re-used or sold as scrap to recover some of the loss. However, if management has never established a policy for salvage control, then this element of loss control is not used to its full, economic advantage.

SUMMARY

A key to the success of any accident investigation and loss control program is the *proper reporting* of all events which lead to a loss. Management must establish a *clear policy* which emphasizes the importance of reporting in the investigation process. If management is unaware of those incidents which downgrade the efficiency of their enterprise, then they will not be able to properly address such occurrences to identify causes and implement corrective actions. The main objective of the accident investigation process is *loss control.* Management can reduce or eliminate the possibility of similar loss events occurring in the future by properly identifying and investigating the cause of each event and implementing actions to preclude the possibility of recurrence. Much of what is now standard industrial safety practice has been developed as a result of previous accidents or incidents which resulted, or could have resulted, in loss. Loss control has become an essential function of most business enterprises since accidental loss of resources (human and equipment/materials) adversely impacts bottom-line profits, both directly and indirectly.

Management is typically responsible for the investigation of accidents (unplanned events which result in some type of loss) as well as incidents (unplanned events which do not but could have resulted in a loss). It is therefore absolutely critical that managers and supervisors become familiar with the nature of accidents, how they occur, and what steps can be taken to reduce or eliminate the potential for their occurrence.

For many years, safety practitioners have understood the simple theory known as the *domino effect* when analyzing accidents and incidents. It is uncommon for a loss event to have resulted from a single cause. More typical is the accident that is the result of numerous downgrading events which occurred in just the right sequence, like dominos falling onto one another, to cause the final loss event. Beginning with a *management lack of control* and ending with the occurrence of the *loss event,* numerous actions will transpire in sequence to downgrade efficiency and increase the loss potential until the final domino falls and a loss occurs. If managers evaluate the stages in which these events occur, they will note that there is usually a *pre-contact, contact,* and *post-contact stage.* They will also discover that ample opportunity for loss control exists in each stage and, if properly utilized, these loss control efforts could make the difference between minor injury and death or minor damage and total equipment loss.

2

Responsibilities in the Investigation Process

INTRODUCTION

In general, the investigation of accidents and incidents is a *management function* and *responsibility*. In the previous chapter it was established that, in most business operations, the nature and origin of those events which become accidents result directly or indirectly from a lack of management control in areas such as personnel, equipment, materials, procedures, and the work environment itself. It follows that management must have complete control over each of these areas in order to ensure the proper control of loss events. An accident in any one of these areas is an indication of a problem, an area that has not been completely examined, understood, or explained to managing supervision. For every undesired event that has occurred, there will no doubt be more than one workable solution to the problems which led to the occurrence. It is true that virtually anyone of intelligence who examines the facts long enough should be able to develop corrective actions that will prevent a recurrence of the loss event. However, it is the management element of an organization that is usually in the *best* position to match a problem with a practical, feasible solution. The nature, type, and extent of the accident does not change this situation, it only dictates the level of management that will be required to conduct the investigation (e.g., first-line supervision, middle management, executive, etc.).

THE MANAGER'S ROLE AS INVESTIGATOR

The role a manager plays in the investigation process, as shown in Figure 2-1, should be viewed as a highly personal one, since managers are ultimately responsible for the events which occur in their assigned areas. Whenever the personnel, equipment, materials, procedures, or work environment in those assigned areas become the subject of an investigation, the manager has a personal interest in the outcome and, therefore, should assume the primary role on any investigation team. If management's interests are entrusted to another party (such as a safety professional or hired consultant) during an investigation, then the manager can no longer maintain control over the investigative parameters or the factors which may influence those parameters. (See *Establishing Investigation Parameters* in Chapter 1.)

When a loss event occurs in a manager's assigned area of responsibility, that manager should know better than anyone else the decisions made, information relied upon, equipment used, and other influencing factors which resulted in the accident. Further, of all those who could possibly perform the accident investigation, only the manager comes closest to fully understanding the capabilities and limitations of the personnel, equipment, materials, and procedures employed in the work environment. In most instances the manager is at the scene when the accident occurs, or shortly thereafter, and usually has first-hand knowledge concerning the pre-contact and post-contact events that took place. The manager knows the peculiarities of each subordinate employee and how each may have played a role in the events which led to the accident. The manager is familiar with the resources necessary to accomplish an investigation and the organizational support requirements essential to the success of each phase of the investigation.

An accident occurs because a problem exists, a problem that was not adequately defined by management, a problem whose consequences were not fully perceived by management, or simply a problem that management did not even recognize as such. A function of the investigation is to determine why information inputs were insufficient to prompt the correct decisions. For the manager, investigation centers on tracing the source of inadequate information and defining the obstacles which may have precluded correct decision-making. Managers must not only define these problems, they must also define solutions. The solution to a problem enhances safety performance, reduces the cost of accidents, and increases the effectiveness and economy of company operations.

Finally, the manager ultimately will be responsible for the resolution of any solutions or actions resulting from the investigation, including the

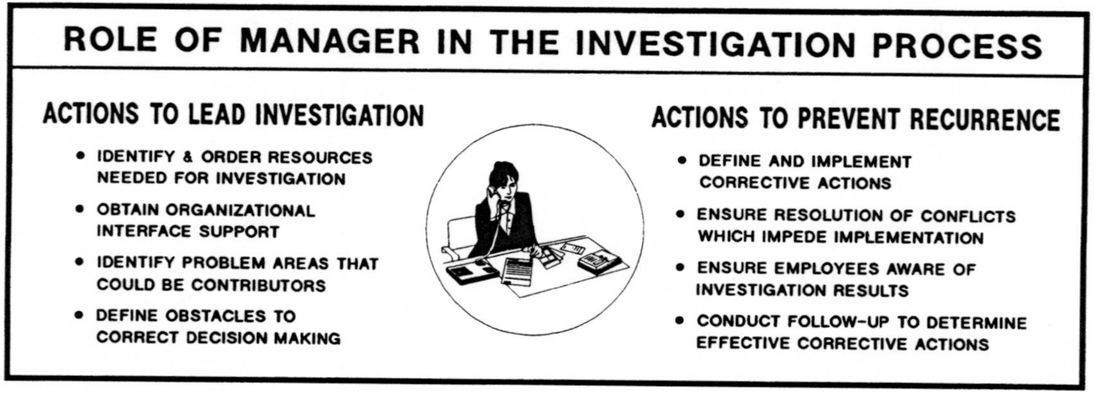

ROLE OF MANAGER IN THE INVESTIGATION PROCESS

ACTIONS TO LEAD INVESTIGATION

- IDENTIFY & ORDER RESOURCES NEEDED FOR INVESTIGATION
- OBTAIN ORGANIZATIONAL INTERFACE SUPPORT
- IDENTIFY PROBLEM AREAS THAT COULD BE CONTRIBUTORS
- DEFINE OBSTACLES TO CORRECT DECISION MAKING

ACTIONS TO PREVENT RECURRENCE

- DEFINE AND IMPLEMENT CORRECTIVE ACTIONS
- ENSURE RESOLUTION OF CONFLICTS WHICH IMPEDE IMPLEMENTATION
- ENSURE EMPLOYEES AWARE OF INVESTIGATION RESULTS
- CONDUCT FOLLOW-UP TO DETERMINE EFFECTIVE CORRECTIVE ACTIONS

FIGURE 2-1.   The role the manager plays in the accident investigation and loss control process.

implementation of corrective actions and follow-up. It is therefore essential that the manager be permitted to perform the investigation to ensure full participation and understanding throughout the entire investigative process. The manager can develop more credible plans and programs suitable for implementation because only the manager fully understands whether the constraints and limitations of certain courses of action are feasible. Managers are best situated to ensure not only that any such actions are started, but also that they are properly completed. Experience has shown that a manager who has been an integral part of the investigation process will undoubtedly be more dedicated and committed to correcting problems than a manager who did not take part in the investigation at all.

In short, if managers are considered responsible and accountable for work performed in their areas of responsibility, then they are also responsible and accountable for any losses which occur in those areas. Few managers would argue with this premise. It should follow, then, that managers be held responsible and accountable for the investigation of those losses and the prevention of any future similar losses in their assigned work areas. The manager must conduct the accident investigation if true loss control is to be achieved.

Additional Aspects of Managerial Investigation

When an accident occurs, the losses suffered are usually irretrievable. Compensatory payments have to be made, damaged equipment has to be repaired or replaced, productivity suffers, and numerous other direct and indirect costs are expended. Although these costs will not be recovered, it is still possible to gain some benefit from a managerial investigation that might not be realized if an organization chooses to utilize a safety professional or hired consultant as the sole investigative authority.

A manager who consistently displays a *positive attitude* toward accident investigation, and reacts promptly and effectively to each such occurrence, demonstrates a sincere concern for the well-being of personnel working in the affected areas. This concern usually translates into employee trust and respect for the manager. Workers not only expect, but actually *need* managers who prove they are concerned with the best interests of their employees. In return, the manager often finds employees who are willing to take extra steps to ensure high-quality work with minimal or no loss potential. It would be impossible to quantify the savings resulting from such mutual respect, but it is obvious that the occurrence of loss events can be greatly avoided simply by fostering an attitude of management concern for employee safety and health. Prompt, conscientious, and impartial investigations identify managers who are in control of their responsibilities. Managers who consistently demonstrate that they can effectively respond to critical situations such as accidents, also reflect an image of managerial capability. It is a known fact that people achieve greater job satisfaction, increased quality and productivity, and lower accident rates when they are confident they are working for someone who has control. Conversely, when employees have little or no confidence or respect for their supervision, morale suffers, productivity decreases, and accident/incident rates begin to rise.

While effective investigation of accidents and implementation of corrective action programs minimize the chance of recurring events, such actions also disrupt productivity and increase costs. Therefore, it must be clearly understood that spending the time and effort to investigate just one accident or incident could prevent work interruptions and subsequent cost increases in the future. Also, during an investigation the manager often uncovers many other loss-potential conditions and can implement corrective actions before any loss occurs.

THE ROLE OF THE SAFETY PROFESSIONAL

Many organizations today have their own safety staffs, whereas many others contract for these services with private consulting firms. Whatever the case, the safety professional will play an important role during the investigation of incidents and accidents. However, it must be emphasized, the safety professional *should not* conduct, control, or direct the investigation. (As stated throughout this text, *the responsibility for conducting the investigation lies with the management of the organization.*) The safety professional should be available to the investigating manager to provide assistance and guidance, as required. For example, the safety professional can help the manager understand specific compliance requirements that may be applicable to the accident situation. He or she can assist in witness interviews, cause determination, and corrective action plan development. But the primary role of the safety professional during the accident investigation is to lend appropriate expertise to the investigating manager and ensure that all possible causal factors and associated findings are adequately researched and documented. In this capacity, the safety professional provides a check-and-balance function throughout the investigative process. Once corrective action plans have been formulated, safety personnel should monitor the implementation of these plans and ensure the initiation of appropriate follow-up by management, when required.

When the extent of a particular accident is such that external regulatory agencies, such as the Federal Occupational Safety and Health Administration (OSHA) or similar state agencies, become interested enough to investigate (as in the case of a fatality), the safety professional should act as the single *point-of-contact (POC)* within the organization. As POC, the safety professional can properly communicate and explain the steps being taken by the organization to investigate the accident, as well as answer any technical inquiries dealing with complex compliance issues that may arise with regard to the accident and/or anticipated corrective actions.

In addition to their role as advisor during the investigation, safety professionals actually have a function *before* the investigation of an accident event. Before an accident occurs, safety professionals must provide competent advice to upper management on investigative policy, accident classification systems, and the importance of pre-planning (including the development of investigation kits and the establishment of training programs). Safety professionals also monitor work operations and activities to ensure investigative plans are up-to-date, with current and changing operational strategies with respect to people, equipment, materials, procedures and the work environment.

Aside from assisting the manager before and during the accident investigation, safety professionals actually contribute to the organization's overall loss control program on a daily basis as well. Safety awareness and incentive programs, safety and health inspections, accident prevention programs, the regular tracking and reporting of injury and illness experience to management, regulatory reviews for applicability, development of safe operating procedures, manuals and plans, and so on, all are activities that effectively contribute to the reduction of loss potential on a continuing basis. In this regard, the manager and the safety professional must consistently work together to ensure the success of the organization's loss control efforts. As summarized in Figure 2-2, the primary role of the safety professional is to work as a partner with the investigating manager during the investigation of an accident event. The role of the safety professional is not minor or trivial, by any means. However, the safety professional should never attempt to control an investigation, which should rightfully be performed by the appropriate management element of the organization.

ESSENTIAL ELEMENTS OF AN INVESTIGATION PROGRAM

It was established early in Chapter 1 that no investigation can occur unless the incident or accident is first reported to appropriate authorities within the organization. This chapter has discussed further the role of the manager as that responsible authority. However, regardless of how conscientious a manager may be, an investigation and loss control program *cannot* succeed

ROLE OF SAFETY PROFESSIONAL IN THE INVESTIGATION PROCESS

- PROVIDE ASSISTANCE & GUIDANCE TO MANAGEMENT
- COORDINATE & ASSIST IN WITNESS INTERVIEWS
- ASSIST IN DEVELOPMENT OF CORRECTIVE ACTIONS
- ACT AS POINT-OF-CONTACT WITH REGULATORY AGENCIES
- DEVELOP SAFE OPERATING PROCEDURES AND PLANS

- ENSURE REGULATORY COMPLIANCE
- EVALUATE CAUSE DETERMINATION
- MONITOR IMPLEMENTATION OF CORRECTIVE ACTIONS
- IMPLEMENT SAFETY AWARENESS PROGRAMS
- MONITOR WORK OPERATIONS TO ENSURE COMPLIANCE

PRIMARY ROLE:

ESTABLISH PARTNERSHIP
WITH INVESTIGATING MANAGER
AND HELP ENSURE THE BEST
POSSIBLE ACCIDENT INVESTIGATION

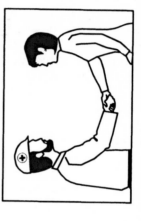

FIGURE 2-2. The role of the safety professional in the accident investigation and loss control process.

without complete reporting. Therefore, beginning with proper and consistent reporting, the following are considered essential elements of an accident investigation program (also summarized in Table 2-1).

Consistent Reporting of All Occurrences to Management

The importance of full reporting cannot be overemphasized. Reporting of accidents and incidents is perhaps the most essential element in a successful accident investigation and loss control program. All employees at all levels must be made consistently aware of its importance from the moment they begin working for an organization. For example, management must implement a clear reporting policy and system and include them during employee orientation and training. There should be frequent reinforcement by management concerning the benefits of reporting and its value to long-term business success. Employees should be confident that a report will receive a positive reaction from management. The experience of sharing accident

TABLE 2-1   The nine essential elements of a successful accident investigation program.

ESSENTIAL ELEMENTS OF AN ACCIDENT INVESTIGATION PROGRAM

1. CONSISTENT REPORTING TO MANAGEMENT

2. INTERVIEW OF WITNESSES AND EXAMINATION OF EVIDENCE AT SCENE

3. DETERMINATION OF IMMEDIATE CAUSAL FACTORS

4. STUDY OF ALL EVIDENCE AND FORMULATION OF INTERPRETATIONS

5. DETERMINATION OF BASIC OR ROOT CAUSAL FACTORS

6. RECONSTRUCTION (IF REQUIRED)

7. ANALYSIS OF BASIC CAUSAL FACTORS AND MANAGEMENT INVOLVEMENT

8. IMPLEMENTATION OF CORRECTIVE ACTIONS

9. FOLLOW-UP

and cause prevention information should create an atmosphere of mutual concern and cooperation.

On-Site Examination of Witnesses and Physical Evidence

Once an accident has been reported, management should conduct an investigation of the accident location and interview those people involved in or witness to the event. It is important that interviews and/or written statements be obtained as soon as possible after the accident. As time passes, especially after dramatic events, people tend to rationalize and analyze what they have seen. If too much time lapses between accident and interview, the investigator may find witness statements flavored with personal opinion, innuendo, insinuation, and the person's own analysis and conclusions as to what caused the accident. It should be noted that it is not necessarily a bad thing for a witness to offer an opinion as to the cause of an accident. However, the opinion should be obtained before the witness has a chance to consider his or her own insecurity, or feelings arising from personal notions of blame or fault. Many times, witnesses may withhold certain facts of evidence if those facts might link them to the cause of an accident. But if statements are recorded as soon as possible after the event, there is less chance that people will have had the time to reason out facts that might make them appear responsible. While many of these concerns might be alleviated with proper training and policy, there is always the chance that a person will not react in a manner dictated by such classroom training.

After obtaining statements from those who were at the scene, the manager should conduct a general appraisal of any equipment and materials involved, procedures in use at the time of the accident, as well as the work environment itself. Each manager should investigate every incident, accident, and hazard to at least this extent.

On-scene investigation and the interview of witnesses often lead to reenactment, as employees attempt to demonstrate to the manager exactly what happened. This can be beneficial at times, since first-hand knowledge of the pre-contact, contact, and post-contact stages might be obtained. However, the investigator must be extremely cautious because reenactment requires establishing and recreating the same conditions which led to the accident in the first place; and, by repeating the same substandard practices, it is entirely possible that an incident can become an accident the second time around. Reenactment should therefore be used only when information cannot be obtained any other way or when it is essential to verify key facts or resolve conflicts in testimony.

[handwritten margin note: Interview witnesses/involved parties as soon as possible so that information will be fresh and unbiased.]

Determination of Immediate Causal Factors
The proper degree of investigation will lead to the identification of immediate causes of the accident (i.e., the symptoms which developed due to unsafe acts and conditions). On-site examination of witnesses and physical evidence will start the investigation process and will help determine the necessity for further reporting and intensive investigation. Upon first review of the physical evidence and an evaluation of all witness statements, the manager should be able to develop a clear picture of those causal factors which directly influenced the outcome of the event. These may include anything from the overused *human error* causal factor to equipment failure and faulty operating procedures. It is extremely important to recognize at this stage that these immediate causal factors are usually only a symptom of some deeper-rooted cause. Basic causal factors are usually linked directly to some failure in the management of the organization. Ending an investigation after the determination of only the immediate causal factors is an all-too-common practice and, subsequently, the primary reason for multiple recurrence of past accident events.

Study of Evidence and Formulation of
Interpretations
In order to complete the evaluation process and ensure the proper identification of the basic or root causal factors, the manager must evaluate all available evidence and develop opinions based on fact. Where specific facts are lacking, the manager should be able to formulate interpretations with some degree of accuracy based upon his or her own background and knowledge of the work environment. A manager knows his or her work area, employees, equipment, materials, etc., better than other people in the organization; personal expertise that can assist in the identification of basic accident cause.

There are numerous methods and techniques used to evaluate and analyze evidence during an accident investigation. These range from a detailed review of all statements, photographs, equipment, procedures, etc., to the performance of a comprehensive *failure modes and effects analysis (FMEA)* or, more commonly, a *fault tree analysis (FTA)*. Whatever the tools used, it is important that an exhaustive evaluation of all available evidence be performed before any determination of root or basic causal factors is made. Anything less would jeopardize the successful outcome of the investigation and increase the risk of accident recurrence.

Determination of Basic or Root Causal Factors
When all evidence has been properly evaluated, the manager should be able to move from the analysis of immediate causal factors to a determination

of basic events which have downgraded the efficiency of the business operation and led to the accident. However, a basic change in philosophy may be required in order to meet this, the ultimate goal of the investigation process. When a manager discovers, for example, that an employee erred because he/she was following a faulty procedure, or equipment failed because its yearly maintenance was allowed to slip due to other management *priorities*, the manager has actually identified failures in the management system that are undoubtedly self-incriminating. Such failures can be insidious and may not be easily associated with accident cause. In fact, it is possible that management decisions (or indecisions), over many months or even years prior to the accident, indirectly affected subsequent decisions that progressed toward the final loss-producing episode. For this reason, proper investigation requires examination of *all* possible factors that could have directly or indirectly influenced an undesired outcome. Consider the domino effect discussed in Chapter 1. As each influencing factor acts upon the next, undesired events begin to occur; if allowed to continue without controls, this chain of undesired events can lead to loss.

Reconstruction, If Required
In some accident cases, determination of basic or root cause may not be entirely possible without more detailed evaluation. In other cases, the manager/investigator may wish to verify certain theories regarding the order in which events transpired. Whether accidents under investigation are particularly complex or relatively straightforward, pinpointing the exact causal factors that led to the loss may require the use of a method of reconstruction. Reconstruction of an accident is an advanced technique used to analyze, through various models, the contact and post-contact stages of an accident. For example, reconstruction may involve reassembling of equipment, materials, and parts of the working environment to demonstrate how each influenced the other in such a way as to result in an accident. Reassembly usually involves supports, frames, molds, and other such devices to ensure structural integrity. Following reassembly, the model is examined for characteristics of failure modes, indications of contact sequence and energy transfer, and any other possible cause-and-effect relationships. Hypotheses can be formed, tested, and substantiated using the model, with laboratory tests of a material's strength and composition supplementing the process.

Scale models can be effective in the reconstruction of an accident or incident occurrence. This method is particularly useful in determining relative positions of people and equipment during the pre-contact, contact, and post-contact stages of the accident event. It is a reconstruction method that is frequently used during the investigation of motor vehicle and aircraft accidents, but it has many industrial applications as well. Such models have proven very useful in developing hypotheses and verifying assumptions. Because the use of models is usually acceptable during legal proceedings, it can also be an invaluable aid to the litigation process.

In some cases, depending upon the exact nature of a particular loss event, mathematical models can be used to compute and evaluate factors such as speed, trajectories, impact force, skids, etc. As with other methods of reconstruction, the mathematical model requires precise measurement of position and size of equipment, fragments, and bodies at the accident site. (Measurement will be discussed further in Chapter 7.)

Analysis of Basic Causal Factors and Management Involvement

Analysis of basic cause determination will become an effective tool only when it is focused on the identification of management deficiencies and not on the superficial symptoms which surface as unsafe acts and conditions. Basic cause analysis traces the origins of the accident to their roots in managerial errors and lack of adequate controls. Management's thoughtful involvement, or lack thereof, in the decisions that affect every aspect of business operations must include consideration of accident prevention and loss control. Lack of such involvement should be evident during the analysis of basic accident causal factors. If management is serious about loss control, then its efforts will be apparent throughout the investigation process. However, even those organizations quite serious about the control of losses can uncover weak areas in their accident prevention program through the proper use of investigation.

Implementation of Corrective Actions and Management Controls

Although all root causes may have been determined, the accident investigation process is not yet complete until management develops and implements controls and other corrective actions that will prevent the occurrence of future similar events. This is the purpose of accident investigation. If nothing is done after causes have been identified, then loss events are sure to repeat at some point in the future. Effective accident investigation yields effective loss control. Each complements the other, but only if management acts decisively with the information obtained during the investigation. Because accidents are often rooted in management's lack of control, it has been clearly established that it is a function of management to investigate accidents, beginning with first-line management. It is also essential that management, at all levels, implement corrective actions to control those losses which downgrade the efficiency of business operations.

Follow-up to Determine Adequacy of Corrective Actions

The only way one will know if a particular corrective action has been successful is to follow up after its implementation to determine its level of effectiveness. A bad plan, or even a good one that has been poorly or incompletely implemented, can result in additional and even more severe incidents and accidents. Measuring the success of an accident investigation and loss control program has never been a particularly easy task. One can only guess how many additional accidents and incidents might occur if accident prevention techniques are not employed. However, there are still some basic measurements which can be taken to show improvement over previous performance. Such improvements might indicate program success. For example, if lost-time injury statistics begin to decline after loss control efforts have been implemented, it might be inferred that these efforts are being somewhat successful. Similarly, if management implements a safety awareness campaign following an incident or accident, to enlighten employees about hazards in their work areas, measurements can be taken to determine whether the level and severity of incidents decrease.

When management can determine, through proper follow-up, that corrective actions are effective in controlling the potential for future loss events, they have made a positive impact on the overall operation of their business enterprise. Follow-up becomes an essential factor in ensuring the long-term success of any accident investigation and loss control program, as well as the success of the business itself.

SUMMARY

This chapter has emphasized *management's role* as accident investigator and presented key elements of a successful investigation program. It must be clearly understood that organizations that consider their managers the ultimate accountable authority for all activities occurring in their assigned areas of responsibility, must also ensure that management remains responsible for the investigation of those events which downgrade operational efficiencies (i.e., incidents and accidents) in those areas. The safety professional *assists* management in these investigations by providing expertise pertaining to codes, regulations, and standard safety practices, but he or she should *never* lead the investigation in place of the manager.

Since *reporting* is one of the most essential elements of the investigation process, management cannot afford to approach the subject of accidents without *commitment* and understanding from employees. Managers must therefore consistently portray a *positive attitude* toward accident investigation that will, in turn, convince employees that their real concern is the

[handwritten notes: positive manager approach to incident/accident reporting = trust from employees = more frequent incident reports made = future accident prevention]

prevention of future loss events that could jeopardize worker safety or increase the risk of lost business opportunities. When such an approach toward accident prevention is evident, employees usually become more concerned with reporting accidents, thus preventing further loss to the organization and themselves.

Finally, when it is either suspected or determined that a manager is a major element of an accident's root cause, it becomes the responsibility of senior management to take charge of the investigation. Clearly, such circumstances warrant upper-management involvement because a lower-line manager may be considered *too close* to the accident to be an impartial investigator. Often, indications that an investigating manager is part of the cause are difficult to recognize; warning signs may develop during the early stages of the investigation. If blame is being sought or assigned rather than causal factors being identified, the value of the investigation should be suspect and the reasons for the mis-focus identified. It is important to remember that when a finger of blame is pointed, three fingers still point back toward the accuser. In such cases, the trained safety professional often becomes the impartial observer who should be able to recognize an investigation that is off-track.

help with the natural response to place blame to re-focus on future accident prevention

3

Investigation Planning and Preparation

INTRODUCTION

Before even the occurrence of an incident or accident, specific plans must be developed to organize the conduct of an investigation in anticipation of any possible future accident, regardless of magnitude or circumstance Only through adequate planning can sufficient resources be available to assure a proper investigation when an accident or incident occurs. For example, investigation planning and preparation requires that management line up capable and trained investigators with expertise in specific areas of concern. If these individuals are identified and properly trained, then the success of any subsequent investigation is much more probable than if investigators are selected and trained after the incident occurs. Other resources, such as investigation equipment kits, should also be considered and obtained during the planning process. An adequately prepared organization will establish procedures designed to preserve fragile evidence from damage or pilferage and designate clear investigation responsibilities well in advance of actual need for such resources.

Every investigation requires the identification and development of related facts so that probable conclusions can be drawn and proven. Planning and preparation assist the investigator in the quick and accurate determination of pertinent facts and the development of valid, practicable conclusions and actions. Although no possibilities should be disregarded without evaluation, proper planning and preparation enables some degree of discrimination between likely and unlikely factors, which is extremely important, especially when time, level of effort, and cost are issues of concern.

Planning results in conservation of investigative efforts and resources and a more effective overall investigation.

Planning and preparation is obviously an important element in the investigative process, whether there is a single investigator or an entire team. However, when a team of experts is assembled to investigate the cause of a particular loss event, good planning becomes even more essential to the success of the investigative effort. When use of a team is dictated, it is usually best for one investigator from the team to concentrate on a specific area of concern, such as the examination of injuries or damages, while another focuses on interviewing witnesses and a third concentrates on evaluation of operating procedures or company records. Others may prepare diagrams and sketches or review past accident/incident history. In a multiple-person investigation team, the team director must pre-plan, coordinate, and allocate resources to meet the objectives of the team as a whole. At regular intervals individual investigators must coordinate their findings with other members of the team to ensure a synergistic evaluation of the causal factors being identified. Proper planning and preparation is the key to the accurate determination of accident/incident causal factors, in the shortest possible time and with the lowest expenditure of resources.

THE PLANNING PROCESS

It is essential to pre-plan investigations so that the appropriate personnel within the organization are familiar with the proper method of reporting accidents to those designated to receive accident facts. The loss control investigative planning process is actually a task-oriented analysis of the required work to be performed before, during, and after (follow-up activities also require pre-planning) the investigation.

A wide variety of activities can occur while planning for an investigation. The level of planning required will be influenced by numerous factors, such as the size of the organization and the complexity of daily operations. However, most experts agree that the process can be described as having four essential steps.

Step 1: Establishing Objectives

In any process, clearly established objectives determine the course and direction of all activities to follow. In an investigation program, the lack of realistic and understood objectives is potentially the greatest threat to the success of the entire program. For these reasons, the investigation planning process must begin with a serious commitment from upper management. Upper management must define the objectives so there is a clear understanding, through all subordinate levels, of upper management's expectations.

Unless management obtains a clear understanding of upper management's expectations, no amount of planning can properly prepare an organization for the investigation process. It is unfortunate that, in the course of daily operations, it is not uncommon to find operational managers and first-line supervisors who re-prioritize general planning activities in favor of production problems and other operating considerations. Even more common is to find that planning tasks have been delegated to lower-level staff. By this time, any plans eventually developed usually end up as books on a shelf that no one reads or evaluates. Then, when an accident does occur, the person who actually may have to direct the investigation (i.e., the manager) will not have the time to become familiar with the written plan. As a result, he or she will usually conduct a less-than-adequate investigation with a focus on time and cost rather than cause and effect. Such an investigation will surely fail to meet the primary goal of preventing future similar events. Therefore, to eliminate this potential for investigation failure, the expected results must be identified by upper management and clearly established in the objectives. However, equally critical, unless upper management's expectations for the results of the investigation are reasonable and achievable, the investigation still will fail. For example, lower-level managers will view a normally reasonable and sound objective (such as *all accidents and incidents will be investigated to prevent recurrence*) as pointless if upper management does not commit the resources necessary to meet that objective.

Successful investigation must have clear and rational expectations.

Step 2: Understanding Priorities

Planning for priority of action in the investigation is considered the next essential step in the process since, without a clear understanding of priorities, there will be no clear method of achieving the objectives established earlier. It is typical during an investigation to find that a myriad of facts and information is being collected and evaluated; much of the information may not appear to have any direct correlation with the intent of the investigation, but must nevertheless be examined. Proper planning for priorities, in advance of the accident situation, will help create a clear path to the objectives. Then, instead of a haphazard trial-and-error method of evaluation, the investigation team can proceed with their charge in a more focused and direct manner.

During investigation planning, management identifies and prioritizes the resources that are to be used for the investigation. A well-defined accident classification system will aid in the process of establishing priorities. For instance, once management has established the various levels of accident classification (such as catastrophic, critical, severe, major, minor, etc.), the planning process can better identify the specific management team members who are to be assigned to accidents of various classifications and also provide for transfer or division of their other responsibilities. Similar provisions can be made for other resources, depending upon the priorities established by upper management. People with the necessary skills required for certain aspects of the investigation, as well as equipment and materials, are also identified and prioritized during the planning process. Improvising with inadequate or inappropriate people and equipment can destroy evidence, damage otherwise repairable items, cause injuries, and damage the inadequate equipment. Proper accident classification and assignment of resource priorities combine to avoid such situations.

It is essential that the reader understand the importance of priorities in the investigation planning process. An inadequate accident plan is one that allocates resources unrealistically, with no clear priority or purpose. When investigation team members are selected, it is often assumed that their normal, everyday responsibilities and activities will be delegated or resolved, thereby freeing them for full-time participation in the investigation. However, unless priorities have been properly established and understood, the assumption that all regular duties will be addressed by someone else often proves erroneous, and both the investigation and normal work are neglected as a result. Such shortcomings only serve to increase the loss arising from the accident as well as reduce the organization's overall responsiveness to future similar events.

Step 3: Development of Actions and Schedules

Once the necessary resources required to meet the established objectives (Step 1) have been identified and prioritized (Step 2), the third step in the loss control investigation planning process requires careful budgeting of the planned resources. This includes the detailed development of alternate courses of actions to be taken in the event certain resources are not available when actually needed (including personnel). Actions should be developed and scheduled for all conceivable types of accidents, especially those the organization has classified as catastrophic, critical, or whatever the nomenclature used to identify the most serious types of accidents. Once such actions have been developed for this classification of accident, then somewhat simplified versions of these plans can be used for lesser classifications. Obviously, the more creative action plans will offer greater options to the manager assigned to perform the investigation. The more the options, the greater the chances for a successful investigation.

As a minimum, proper planning should address the numerous repetitive actions that are normally required for all investigations (such as the notification process, response actions, area controls, witness interviews, the collection and preservation of evidence, etc.), so that such actions will not have to be thought out each time an accident occurs. Proper planning should include preparation for unique accident situations requiring specialized activity or equipment (hazardous materials spill-response equipment, fire/emergency equipment, special lock-out requirements for certain machinery, etc.)—In general, planning in advance of any foreseeable accident.

Another important aspect of proper action planning is to ensure the protection of the investigators themselves against any possible physical injury and/or hazards to health that may occur during the investigation process. In fact, without proper action planning, recurrence of the subject accident (including injuries to personnel) could result, with the investigators as unintended victims.

Detailed action planning should also include consideration of report scheduling and information transfer. The method, frequency, and degree of information reporting can be a critical aspect of investigation success. Interim reports to upper management, progress reports to other investigation team members, and timely communication of investigation progress throughout the organization and to interested external parties, all are examples of the type and degree of reporting that should be carefully considered and scheduled during the planning process. When interested parties are provided with accurate information they may need to perform their tasks effectively, there is less of a chance that their actions will adversely influence or affect crucial evidence or the progress of the overall investigation itself. An adequate and thought-out provision for communication and information transfer will help ensure the smooth flow of investigative activities.

A detailed plan of action, developed by those managers most likely to be affected by such actions, should schedule the expected, required, and anticipated activities, as well as provide the necessary flexibility to accomplish systematically those actions that must, should or might be performed during the course of the investigation. The plan should identify in advance and develop acceptable provisions for those aspects of the investigation that, if delayed or neglected, could prove most troublesome to the investigators. Plans also provide investigators with appropriate access to sources of assistance by establishing authority and identifying organizational capabilities that will be made available to the investigator, should a need for such arise during the investigation.

Finally, action plans should provide for an orderly process of return to normal operating activities. It is important for planners to realize that the sooner the organization can resume its normal, day-to-day activities, the better the chance of reduced loss as a result of the accident event. The longer it takes to investigate an accident, the greater the amount of time, effort, and resources lost as a result. This is not to suggest that an investigation should be rushed through to save money. However, proper scheduling of investigation activities does provide certain control and establishes specific milestones that must be accomplished in order to properly complete the investigation in a timely manner.

Step 4: Preparation of Corrective Action Procedures
The final step in the planning process concerns the development of procedures that will be used by the investigation team. Adequate provision of operating procedures completes the planning process by describing how the *courses of action* identified in Step 3 will utilize the *resources prioritized* in Step 2 to accomplish the upper management *objectives* established in Step 1. A thorough effort to develop procedures may also yield an additional benefit: this aspect of the planning process often identifies basic program deficiencies that, if corrected, actually result in the prevention of accidents. Another attractive aspect of procedure development during the planning process is flexibility. Procedures can be in the form of short checklists or placards, amplified checklists, instructional notebooks, guides, manuals, detailed operating procedures, or any other form that facilitates their use.

Once procedures have been developed, all possible members of an investigation team (including all members of management) must become thoroughly familiar with the contents of each procedure so that there will be no unexpected surprises should an investigation team be called at some future time. Some individual section managers may wish to add to or modify a specific procedure that affects their area of operation, to improve clarity and decrease the likelihood of misinterpretation during an actual investigation. However, individual managers should never attempt to alter drastically the intent of the written procedure without first obtaining concurrence from peer management team members as well as upper management. Preferably, all managers who might be affected by written procedures should be allowed to participate in its initial development so that no real alterations will be necessary after the document has been approved and released.

Each of the four steps to the planning process described above should exist, in one form or another, in every successful accident investigation and loss control program. Without a clear, planned, systematic approach to the investigation of loss events, organizations will not gain adequate control over their losses resulting from repetitive accident occurrence. It is poor business sense to view such events as *tough luck,* or simply to accept the myth that *accidents just happen, and there is nothing that can be done about them.* Proper planning not only helps to prepare an enterprise for appropriate actions in the event of an accident, it often identifies potential loss factors that can then be corrected or controlled before an accident occurs. The truth of the matter is that accidents do not *just happen,* they are *caused*—and *planning* is something that can be done to prevent them.

The Initial Phases of Accident Investigation and Loss Control

Having established and discussed the four steps in the planning process, the following describes the importance of proper planning in each *phase* of accident investigation and loss control. The four steps of the planning process, discussed earlier, are the common denominator of the investigation phases described below.

Accident investigation and loss control occur as a set of organized and planned activities. These activities can range from the very simple, such as the recording of times, dates, and participant information, to the highly complex, such as performing microscopic metallurgical analyses to determine stress, tension, and failure points. To facilitate the organization of these many and varied activities, the loss control planning process should include some consideration of the order in which these activities should occur. By dividing the process into *phases* of activity, the investigation team can better organize necessary actions and events, as required, to reduce potential losses to the lowest possible level. Through systematic evaluation of the activities that occur or should occur in each phase, the investigating team can also identify shortcomings in the entire planning process that contribute to the loss potential of any given accident situation. Generically speaking, accident investigations and loss control activities can be divided into six distinctive phases.

Phase 1: Before the Occurrence

What investigative activities can one plan *before* an accident event occurs? There are, in fact, numerous loss control actions that should be undertaken prior to an event, some of which might indicate possible shortcomings in a management system. On the other hand, the absence of proper planning actions *prior* to the occurrence of an accident can seriously affect the magnitude of the final outcome of a loss event. This is not to suggest that accidents themselves should be planned events. After all, it was established in previous chapters that an accident is, in fact, an *unplanned* event. However, it is entirely possible to *reduce the potential for loss* resulting from an accident through simple consideration of any actions that, taken before an accident occurs, might mean the difference between an inconvenient but temporary shutdown of operations and a total loss of business and employees. Indeed, accidents themselves are not planned events, but *preparation* for the control of potential losses resulting from possible accidents is essential for the success of an accident investigation and loss control program. Some possible pre-accident loss control actions that should be implemented include the following suggestions:

1. Management should consistently provide appropriate training, communication, and information to employees throughout all levels of the organization concerning the potential for accidents. This will not only stimulate attentiveness to accident prevention, it may also reduce losses that could occur during the post-contact stage. If personnel are made aware of accident potential, they may be more cautious during the performance of their everyday duties.

2. Coordination and regular communication with external response agencies—fire, medical, police, utilities (gas, water, electrical), bomb disposal, poison control, etc.—will serve to familiarize such agencies with the potential for accident situations and will also ensure a quicker, more effective response, if an accident should occur. The latter might prove instrumental in keeping any post-contact losses to a minimum.

3. Secondary accidents frequently increase the severity of loss or endanger rescue workers and investigators. Potential sources of secondary accidents should be identified in any operating plans or procedures, job orientation, group communications, etc., so that employees are aware of such possibilities. In many cases, the threat of secondary accidents can be diminished through simple consideration of accident potential before a loss event occurs. Likely sources of secondary accidents include the nearby storage of gas or liquid fuels, combustibles, toxic agents, caustic or contaminating materials, electrical equipment, etc.; these could be isolated from one another in the planning stages so that the chance of secondary loss is minimized. The loss control investigation should focus on the organization's consideration of secondary accident potential and the adequacy of any steps taken to reduce or eliminate such potential.

4. Losses resulting from accident events can also be reduced if consideration of their occurrence is provided for in collective bargaining agreements, insurance contracts, delivery schedules, or any other contractual agreements that might obligate a business to continue payments or expenditures even during periods when operations are interrupted due to the occurrence of an accident. It is not uncommon to find union agreements that provide for complete or partial compensation to represented employees when the employer has ceased production due to circumstances beyond its control. A simple provision in any such agreements might be difficult to negotiate but could prove critical in the reduction in losses incurred after an accident.

5. Any accident response kits, special tools, and/or equipment that will be used during the investigation of an accident should be planned for well in advance and made readily available at all times. Equipment or instruments necessary for securing an accident scene, or that required for the proper recording, removing, preserving, and examining of evidence should be considered during the planning process. If such provisions must be hurriedly assembled after an accident has occurred, precious time will be lost which could contribute to an increase in loss potential. Loss control accident investigations should evaluate the planning process to verify an organization's readiness to respond to such events.

Phase 2: Discovery of the Accident Situation

Once an accident has occurred, there are specific activities that should take place and that could still have an effect on the resulting degree of loss incurred. The planning process should identify and prepare for the accomplishment of these activities. Any subsequent accident investigation should evaluate the adequacy of these actions to ensure that all possible attempts to reduce loss have been taken. Aside from the obvious response actions necessary to prevent further immediate loss, an organization should also be prepared to accomplish at least the following after discovering that an accident has occurred:

1. Management personnel must be notified of the occurrence as soon as possible. The level of management notification should be commensurate with the degree or classification of accident severity. Such provisions should be clearly established so that no question will remain concerning the level of management notification. Proper notification of essential management representatives will permit the prompt execution of critical decisions that might affect the degree of loss resulting from the accident. For example, a simple telephone notification list (current and up-to-date), provided to shift supervisors, switchboard operators, security guards, or whomever, might be an organization's focal point of communications during an emergency and will expedite the notification process. Investigators should verify the existence of such provisions and evaluate their adequacy in terms of maximum loss control potential.
2. Depending upon the nature and circumstance of a given accident situation, there may also be requirements for the immediate or timely notification of certain state and federal regulatory agencies. For example, the U. S. Occupational Safety and Health Administration (OSHA) must be notified in the event of a fatality or the hospitalization of a specified number of employees. The Environmental Protection Agency (EPA) and associated state and federal support organizations and agencies must

be informed of accidents that adversely impact the surrounding environment to a specified severity and extent. The planning process should include provisions for such notifications so that any penalties, fines, or citations for non-compliance with reporting requirements can be avoided. Such fines, which only increase the degree of loss resulting from an accident, are completely avoidable through proper planning.

3. Public relations and communication of factual information to the press is an extremely important element of the planning process. Depending upon the severity of a particular accident or the geographical location of the accident scene relative to surrounding (public) areas, organizations should ensure provisions for the release of accurate information to interested members of the press or other media. Organizations that do not consider the long-term damage that can be mercilessly inflicted on their business operations by the media will surely fall victim to its own shortsightedness. The press and media can be cooperative and useful allies in an emergency situation by providing a quick means of distributing vital information to the public. However, when public relations and communications are handled poorly, news agencies can become fierce enemies of an organization they feel is withholding information or attempting to keep the public uninformed. Management should consider the damaging effect of faulty or inaccurate press coverage on long-term profits. They should plan appropriately for such possibilities by ensuring designated and *trained* company representatives will be available to communicate, as accurately as possible, information to the press during times of emergency. Such provisions could help drastically reduce the potential for continued social and economic losses long after the accident has occurred. It is important that those who are authorized to make statements to the press on behalf of an organization are made fully aware that what they say can affect pecuniary liability or community and labor relations. Official statements made to the press should be carefully structured and worded so as to provide the maximum amount of information possible without jeopardizing the company's best interests in the process. When evaluating the effectiveness of an organization's loss control efforts, investigators should examine any public relations program that might or should be in place. A more detailed discussion of the issue of public relations is provided in Chapter 4 of this text.

Phase 3: Arrival at the Accident Scene

Planning for some specific activities, which should occur once responders reach the accident site, can also prove effective in reducing the value of the ultimate loss resulting from the accident. These may include any of the following recommended considerations:

1. Investigators should verify that all appropriate steps have been taken to ensure the prompt and effective rescue of any injured or endangered personnel in such a manner that additional injuries are not inflicted or existing ones compounded. Planning should address the provision for quick first-aid treatment, the ability to accurately assess other potential medical factors such as the need for professional medical personnel, as well as the capability to provide for temporary physical and psychological comfort to those who were immediately involved in the accident event. Organizations that plan well for the worst possible outcome of any accident situation will fair the best during an actual loss situation and, as a result, the total subsequent loss will be kept to an absolute minimum. It therefore becomes an essential task of the investigator to evaluate the organization's overall abilities to respond to any and all possible accident events.

2. Planning should prescribe procedures designed to reduce the risk of any further injury or damage in order to effect maximum control of losses resulting from the primary accident event. For example, procedures should be developed to neutralize flammable, acidic, caustic, toxic, or radioactive energy sources and to prevent personnel contact with such hazards. Organizations should be prepared, with contingency plans designed to reduce risks such as those associated with pressure vessels that might rupture, move, or suddenly discharge their contents, or the hazards of a weakened or unstable structure that could collapse. Safeguards against the ignition of flammables or explosives, ventilation of toxic or asphyxiating atmospheres, and protection against suffocation and drowning are examples of other loss-reducing measures that can be assured through proper planning.

3. Personnel arriving first at the scene of an accident have the advantage of viewing it in its natural state, before anything has been touched, moved, or displaced. Planned actions for emergency response teams, management representatives, and workers should include observation of people, equipment, materials, and the work environment. Immediate attempts should be made to identify those obvious elements that may have been or were involved in the accident occurrence. The astute observer, with the ability quickly to assess an abnormal situation, will actually facilitate the beginning of the investigation process by identifying those aspects of the accident scene that do not appear to be right or normal. This action will also help initiate the reporting process since abnormal conditions must be documented and appropriate management officials immediately informed. Inadequate or improper first-response activities could add significantly to the overall loss resulting from the accident. The investigation should therefore examine the organization's

readiness for first-response and ensure all necessary considerations have been evaluated during the planning process.

4. Secondary to, but equally as important as emergency actions are those activities necessary to preserve the physical and intangible evidence from loss or distortion. For example, planning should include provisions for the removal of extraneous personnel (observers, bystanders, curiosity-seekers, etc.) from the accident area. There should also be established a procedure for the stoppage of any additional work to prevent the possible disturbance of evidence. In all cases, a great deal can be learned about accident cause through the comprehensive evaluation of all available evidence. Unfortunately, this may include an analysis of any human bodies or body parts that may have been a result of the accident event. It is therefore critical to the preservation of evidence that accident planning include a prepared permission form to obtain next-of-kin approval to conduct an autopsy, with a designated person responsible for contacting relatives and obtaining the required permission. Although such actions may be somewhat uncomfortable to consider, they are much easier to perform *before* an accident occurs than after. Time will be saved, and therefore losses reduced, if these unpleasant duties are considered well in advance of any actual need.

Phase 4: Gaining Control of Emergency Conditions

Once the accident situation and any resulting emergency conditions are under the control of an established authority, such as an investigation team or other designated management representatives, there are additional activities which should occur that will contribute even further to loss reduction. Adequate planning in each of these areas could effectuate a decrease in the total potential loss:

1. Depending upon the nature of a particular accident event, there may be requirements under certain federal, state, or local laws for external agencies to conduct their own investigation. Public laws may require state or federal safety and health agencies, environmental agencies, transportation agencies, or even local police and fire departments to conduct an independent investigation of the accident event. For planning purposes, organizations should obtain appropriate legal counsel to consider just how such investigations will affect the company's own internal investigation process. Just because external entities may be entitled by law to investigate, such agencies are not authorized an exclusive right of access to company property and facilities. Plans for the coordination of such investigations must be well established during the planning process. Where the possibility exists for more than one external agency investi-

gation, provisions should be made for the protection of evidence of interest to each agency and procedures written addressing both joint and separate examination of evidence by such interested parties. Planning should therefore include meetings with all potential investigating agencies well in advance of any accident situation. Such action will help establish mutually beneficial understandings and might avoid the possibility of charges of obstruction and subsequent fines resulting from external agency investigations.

2. In most accident cases, especially complex ones, there is often a requirement for the systematic collection and examination of all possible sources of evidence. During the planning process, the organization should include provisions that take into consideration the degree to which such evidence is susceptible to further loss or distortion. Any special tools that may be needed to disassemble damaged or apparently deficient equipment should be identified and acquired well in advance of actual need. Also, if personnel with specialized training and qualifications will be required to operate such equipment, then they must be identified and trained on the nature of accident investigation before the occurrence of a loss event requiring their services. Planning should include the identification of the source of any special equipment and the personnel needed to operate it; definition of the investigation parameters that will justify the diversion of these resources from their normal assigned duties; and the method by which the equipment and personnel will be transported to the accident site. Poor planning in this area could seriously jeopardize the preservation and examination of all available evidence and increase the risk of conducting an unsuccessful investigation (i.e., one that fails to properly identify and evaluate all causal factors).

3. Once the accident situation is under proper control, but before the actual investigation begins, investigators should be refreshed on the proper methods and techniques of the investigation process, as well as briefed on any peculiarities that may be associated with the subject accident event. In addition, they should be allowed to review plans or procedures on a regular basis to ensure complete familiarity with established protocol. When more than one person is to enter an accident investigation, they should receive a thorough indoctrination. When investigators do not remain current on the investigation process, or are not properly briefed on the specifics of a particular accident prior to participation in the investigation, they may have a tendency toward assumption, speculation and conjecture. Assumptions can cause the loss of valuable evidence or injuries to investigation team members and other personnel.

4. After gaining control of an accident situation, decisions may have to be made regarding the temporary or complete shutdown of facility opera-

tions. If proper planning has occurred before the accident, there should be provisions in all applicable company contracts to allow for the suspension of work, call-back approval authority and priority, return-to-work job instructions, and procedures for keeping suspended employees informed of investigation status. Methods of implementing and enforcing certain restrictions on work areas or activities, until normal work can be resumed, should also be spelled out in advance. Even the priority for reopening areas such as toilets, locker rooms, food services, etc., should be established. These are all potential sources of additional loss to an organization and, without proper consideration during planning, can be extremely detrimental to the overall success of the loss control process.

Phase 5: After Completion of Data

Once all pertinent data has been obtained and a detailed analysis of the accident facts is about to begin, there are some important investigative aspects associated with data analysis that require planning to ensure the maximum benefit is derived from this effort. Investigators should evaluate accident planning for at least the following considerations:

1. Many investigations might require additional expertise for the examination and testing of evidence. For example, precise questions on material composition, stress and structural activity, ergonomics and other human psychological and physical factors, are often beyond what can be realistically expected of the manager or investigator, the safety engineer or advisor, or any of a firm's other investigative resources. If the organization waits until after an accident has occurred to identify available sources for such technical expertise, serious delays in the completion of the investigation will result in a subsequent increase in loss potential.

2. Depending upon the nature and circumstance of a particular accident event, it may be necessary to communicate certain information pertaining to product liability to external parties such as suppliers of equipment and materials. The legal aspects associated with access to evidence must be examined so that the right to access is not inadvertently denied to such parties, thereby compromising any subsequent claims. If a requirement exists for technical assistance in the examination and allied defense against any liability claim, then organizations should plan for such and be prepared to assist. An additional loss control benefit can be realized by ensuring advanced knowledge of these provisions during the planning process, which may actually influence the quality control efforts of suppliers, thereby decreasing the loss potential during the pre-contact phase (purchasing). By properly preparing for the likelihood of litiga-

tion, and taking prompt action as soon as practical after an accident occurs, the organization can reduce the loss potential dramatically.

3. Once the collection of important data has been completed, there must be provisions for the removal and replacement of damaged equipment and environmental structures. It is strongly recommended that a sequence of phased removal be planned that will better serve the investigation process. After a specific item or work area has been evaluated, debris removal will enable investigators to move on to the next area by eliminating obstructions caused by materials. Planning for the removal of heavy obstacles includes ensuring that the equipment and operators needed to accomplish such a task be identified and readily able to assist. Phased removal of accident debris will also eliminate, or at least reduce, the operational pressures to expedite the investigation by demonstrating highly visible signs of progress. Programming and scheduling provide the required special resources and support while reducing costs and subsequent losses. Plans should also consider the safety of personnel and equipment, as well as quality assurance before any removal of lockouts, temporary construction or installation is allowed to occur.

Phase 6: Additional Activities

Once all work at the accident location has been completed, there are still some activities that must be considered during the planning process which, if properly thought out, could result in a further reduction in the loss potential. For example:

1. Accidents often result in the generation of varying degrees of physical evidence, such as damaged equipment, materials, and debris. It is not uncommon following an investigation to have to re-examine evidence at some point in the future. Planning for the temporary storage of evidence will not only aid in its preservation, it may also reduce the possibility of any adverse impact on other activities. Retention of evidence avoids the likelihood of gaps in the analysis by providing for re-examination when necessary. Organizations must therefore recognize the possibility that secure storage space may be required, and to plan for such accordingly. As with all other aspects of the planning process, having to locate and secure required space *after* an accident can cause significant delays in the investigation, which will inevitably increase the accident's resulting loss.

2. When the investigation is closed, the causal factors identified, and recommended corrective actions planned or implemented, the final disposition of the evidence must be considered. There should be plans for detailed inspection of any damaged equipment or materials to determine the usefulness, serviceability, economic repair, or salvage to avoid unnec-

essary loss of usable items. These actions will also help preclude the possibility of any future accidents or losses by preventing unserviceable equipment from re-entering the work environment.

3. Another important aspect of the overall investigative effort is the provision of adequate facilities for use by the investigation team to conduct interviews, prepare reports, analyze data, and perform other general team activities. Organizations that plan for these resources avoid the waste of premature allocation and decrease the loss potential resulting from the accident. Management should also consider ensuring appropriate resources are available should any follow-up investigations become necessary.

4. Before normal activities are permitted to resume after an accident, there should be a detailed inspection of the subject work area(s) and related equipment to assure appropriate safety and quality controls are in place to prevent any subsequent or additional losses.

INVESTIGATOR SELECTION AND TRAINING

An extremely critical element of the planning process is the selection and proper training of the investigation team members. Every person in an organization who participates in the gathering of facts associated with an accident is, in effect, an investigator. This presumption applies from the first-line supervisor to the senior executive manager attempting to identify the cause of minor or major injury or damage to property. Virtually anyone who may take part in the search for facts and answers is an investigator. Some may already possess the necessary qualifications to investigate accidents due to their technical and professional expertise or experience in the investigation of losses. Others qualify as investigators based upon their highly refined special skills, knowledge, or capabilities in particular areas such as human factors or metallurgy. Many organizations have other personnel on staff who are particularly proficient in operational or maintenance tasks, or who have broad knowledge of day-to-day factors in the occupational environment that often become the basis for accident events. Line managers are also useful in an investigation because of their special knowledge of the work environment and the people, equipment, procedures, and materials used in their assigned areas of responsibility. Executive managers have the required skills needed to coordinate, direct, and control the actions of other investigators.

Regardless of the size of an organization or the degree of loss (or potential loss) associated with a particular accident or incident, the attributes of potential investigation team members should be examined during the planning process. Their roles in the investigation process should be clearly defined, and any necessary training should be provided in advance to prepare each for the responsibilities of investigator.

The number and type of investigators assigned to any investigation will depend on numerous considerations, among them:

- specific objectives, policies, investigative parameters, and loss classification system established by the organization;
- size and nature of the loss incurred;
- actual and potential severity of the loss;
- potential for recurrence of the loss event.

In addition, each of these factors will be affected by the particular nature of the business enterprise and the specific priorities established by the organization's upper management. These considerations should be carefully evaluated to determine whether the investigation can be accomplished by one individual, or by a small group of highly skilled but specialized investigators, or by large teams with sub-teams of people performing individually assigned tasks within the overall investigation.

Investigator Selection

The selection of people to participate in investigations is initially accomplished with the establishment of the organization's accident classification system during the planning process. When an organization defines the various classes of accidental loss (e.g., catastrophic, critical, severe, serious, minor, etc.), they have not only identified the severity of actual or potential loss, they have also established the framework for identifying the necessary resources in people and equipment required to investigate the loss. For example, an accident event with the potential for total and complete loss of an entire facility or the fatality of even one employee would warrant the participation of the most senior level manager in the organization. Other members of such a team might also include plant or facility managers, maintenance managers, safety professionals, and other specialists. Whereas, a minor incident may only require an investigation by the responsible supervisor within the work area. Figure 3-1 shows the organization of a hypothetical investigation team for a minor accident, while Figure 3-2 provides an example of a team structure which might be formed to investigate a critical or catastrophic loss event. In each case, it should be noted that the safety professional participates in the investigation as an advisor to the team and, primarily, to its director. Safety professionals can provide technical guidance on the investigation process and optimum use of resources during the investigation. These investigation team structures are only suggested to demonstrate the necessary flexibility of team composition, which

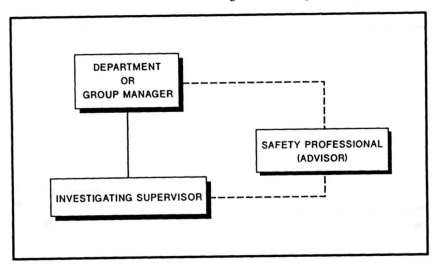

FIGURE 3-1.   Suggested organization of a hypothetical investigation team for a minor accident event.

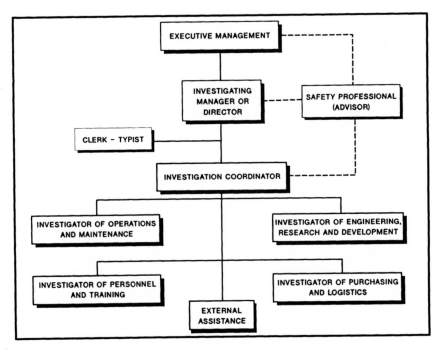

FIGURE 3-2.   An example of a team structure that may be formed to investigate a critical or catastrophic loss event.

must be considered during the planning process; the *team concept* is the means by which the organization will return to normal work operations. Whether the investigation is to be performed by one person or a large group of individuals, the *team* functions that are performed provide for systematic control of the operational chaos which usually follows an accident event.

There are no clearly defined criteria to ensure the proper selection of investigator team members, since each organization will have different needs, objectives, policies, procedures, etc., all of which affect, to some degree, the decision even to investigate loss events. It is noted that investigators are first selected based upon their individual knowledge and experience in a given subject area. However, as a basic guide, knowledgeable individuals who possess certain fundamental *human qualities* should be chosen to investigate accidents on behalf of a business enterprise; these qualities include such unmeasurable traits as integrity, objectivity, perseverance, curiosity, imagination, humility, intuition, and tact, as well as the ability to remain observant under varying degrees of stress. Such qualities must be developed by investigators, rather than investigators selected based solely upon their possession of these characteristics. It is *knowledge first* and *qualities second* that determine the adequacy of a potential investigator. In most cases, excellent mechanical skills, coupled with exceptional oral and written communications skills, are also essential characteristics of a successful investigator.

In some cases, especially when accidents involve particularly complex circumstances, organizations may often resort to external assistance from specialist investigators. It is uncommon to find a business enterprise that contains every conceivable type of expertise within its current personnel base. Depending upon the nature of a loss event, organizations may seek assistance from external specialists in the medical, psychological, legal, or technical arenas. Such assistance can be obtained through consultation with governmental offices such as the National Bureau of Standards (NBS), or located by contacting professional trade organizations such as the American Society of Safety Engineers (ASSE). Numerous consultant directories are available, such as *Best's Safety Directory,* where expert witness assistance can be located.

Regardless of the size or composition of the investigation effort, the provision for some degree of training for those persons selected as investigators should never be overlooked.

Investigator Training

Based upon all the factors discussed thus far, it should be fairly obvious that accident and incident investigation can become a rather complex task, even when loss events are considered to be minor. The necessity for training or, at the very least, orientation for those involved in the investigative process should also be readily apparent at this point.

The investigation of accidents and incidents will usually involve simultaneous but independent performance of numerous specialized functions and tasks. These may include the conducting of witness interviews, taking of photographs, the mapping or sketching of the accident site, examination of records, the identification and tagging of parts, the safe removal of damaged equipment and materials, reaction to inquiries from the public and news media, litigation, and interaction with upper management. The assumption that any one potential investigator, such as a supervisor or manager, can possess all the necessary skills to properly accomplish each of the tasks without some degree of preparatory training, is absurd and will certainly set the investigation up for failure. The pressures placed upon investigators to identify the facts, define the causes, recommend and implement solutions, and return to normal operating status, dictates the necessity for preparation and training. Skill training also becomes essential to avoid the inadvertent destruction of evidence during the investigation process. Efficient and effective investigation results when properly selected investigators are provided adequate resources and the necessary training in investigative duties, procedures, and techniques.

Therefore, a formalized training program becomes a critical element in the accident investigation plan for any organization. All personnel who may be selected as investigators must be trained on the purpose, discipline, and methods essential to the investigation process. In general, the background information and technical instruction should at least include the following:

1. organizational policy, purpose, and philosophy regarding the investigation of accidents and incidents;
2. standards and guidelines established for the development of investigations and the compiling of reports;
3. causes of loss events, and the controls which can or should be used to prevent them;
4. equipment and materials required to aid in an investigation, including a description of the resources available within the organization and the required priorities and procedures used to obtain them;
5. proper response actions to be taken by whom, at what time, and for how long as well as the specific responsibilities assigned to each responding party;
6. interviewing of witnesses, the recording of statements, and evaluation strategies associated with each;

7. techniques pertaining to the proper collection and examination of evidence as pertains to each investigative function;
8. proper methods of communicating with external investigative agencies and the importance of ensuring the confidentiality of specific company information while maintaining a professional relationship with these agencies;
9. interrelationships with other members of the investigative team;
10. sources and types of special assistance that will be made available, if required;
11. public relations techniques, interaction, and communication to ensure sensitivity to the potential impact of statements made to the news media, representatives of parties which may seek litigation for injuries/damage, and for the good of the public;
12. analytical techniques to determine causal factors;
13. proper methods associated with the reenactment of accident events; and,
14. closing of the investigation, which may also include training on the importance of proper and adequate follow-up activities once the investigation has ended and recommended solutions/corrective actions have been implemented.

These basic elements of the training process should generally apply to every investigator, regardless of their level of responsibility within the organization (i.e., first-line supervisor through executive).

Once the level, degree, nature, and specificity of the training program has been established, the organization should also consider the use of simulation as a potential training tool for investigators. When each trainee has been adequately indoctrinated and briefed on the general factors discussed above, it is often quite useful to conduct an exercise involving a simulated accident, which must be evaluated for cause by the newly trained members of the organization's investigation team. This process provides a useful transition between the theory and philosophy provided to the investigators during their instructional training, and the actual practice of accident response.

Simulation is an efficient, low-loss (i.e., low-cost) method of enhancing management training. It not only establishes hypothetical loss events, it also provides a method of performing an intensified examination of a particular work activity or environment, identifying weak areas in the work environment before an accident or incident occurs. To the newly trained investigative eye, normal work practices may be viewed from quite a different perspective than the usual. Practicing witness interviews with work area personnel can provide a detailed analysis of job tasks, as well as test an employee's interrelationships with other workers in the area. Shortcomings in training and qualifications may also be evident during a practice interview. Simulating the examination and evaluation of evidence allows for additional quality control inspection of equipment and materials. Since these are all normal management functions anyway, the training of managers as investigators should be performed as a routine function of the management evaluation process.

SUMMARY

The comprehensive investigation of accidents and incidents is essential to controlling losses and ensuring the overall success of any business operation. But successful investigation will never occur without thoughtful planning and preparation. Management must establish clear *objectives* so that the entire organization will have the same understanding of company *priorities* pertaining to loss control. Once these basic elements of the planning process have been developed, specific *actions* to be taken in the event of a loss can then be defined, with appropriate assignment of investigative duties to individual management team members (i.e., the investigators).

Proper planning for the reduction of loss not only requires consideration of actions to be taken *after* an accident or incident has occurred, but also establishes requirements for action that can be taken *prior* to the occurrence of an accident. This, of course, is a much more difficult proposition. But ensuring maximum loss control at minimal cost mandates both pre- and post-accident planning.

Another critical element of the planning process involves the selection and training of investigation team members. Waiting until an accident has occurred, before assigning investigation duties, will only jeopardize the successful outcome of the investigation and prove counterproductive to the organization's entire loss control program. Preparatory training for investigators is absolutely necessary to orient them to the precise nature of the investigation process and ensure the maximum benefit to the organization. There are a variety of proven, low-cost training methods such as *simulation* that prepare investigators for the responsibilities associated with loss control.

In its selection of investigators, management should focus upon individual knowledge and expertise and how each can be applied to the investigation process. Once these capabilities have been identified, there are also certain personal attributes which would further enhance the value of an investigator's contribution. These include innate qualities such as integrity, objectivity, perseverance, curiosity, observation, imagination, fairness, humility, tact, and intuition.

Investigations must be adequately *organized* and *managed* in order to succeed. Therefore, it is essential that each investigation be controlled by designated representatives of *management*. Safety professionals, external consultants, and technical experts make a definite contribution and provide an invaluable service through their discipline and expertise. But it is management that must actually organize and coordinate the overall investigation.

4

Accident Response Actions

INTRODUCTION

The overall theme of this text thus far has focused on the importance of accident investigation in the *control of losses*. An essential element of the investigation and loss control process is the organization's initial *response* to accident events. It has been established previously that the method used to report accidents, how those reports are disseminated through the organization, the initial-response actions taken to control activities at the accident scene, the actions taken by workers in the area, as well as those taken by emergency response teams, all will influence the level and degree of loss incurred as a result of the accident event. With the exception of response activities, the value of proper planning for loss control has also been discussed thoroughly. This chapter will expand on the importance of *accident-response* actions and examine the value of *response planning* as part of the accident investigation process.

To ensure response readiness, an organization must first have a clear understanding of its own accident risk and loss potential. This understanding is achieved through a detailed analysis of company operations and an evaluation of the risks associated with the performance of those operations. If management can outline the nature and degree of potential loss, based upon the hypothetical occurrence of the worst possible accident in each work task (sometimes referred to as the *maximum credible event)*, they can better plan the appropriate response actions that will reduce the overall resultant loss. Planning for the worst possible accident allows for consideration of all potential response actions that might be required. Then, even if an accident occurs with only minor or negligible risk of loss, the organization is prepared to respond in an appropriate manner to ensure the minimization of any further loss. Since valuable information can be lost if an investigation is not started promptly, and the extent of the loss may increase if proper actions are not taken, it is in a company's best interest to give considerable attention to accident response. Quick and proper accident response demonstrates effective loss control.

RESPONSE ACTIVITIES

Once an accident has occurred, management has numerous opportunities in its response activities to decrease loss potential simply by planning. Accident investigators should examine management's response to determine if all possible loss control actions were taken in a timely fashion. Beginning with the initial *notification* of an accident event, management can effectively control impending losses by executing a planned course of action.

Notification Actions

The initial notification of an accident event can be generated by a wide variety of sources, depending upon the scope and dispersion of business activities. For example, an accident occurring in an industrial area might be reported by a first-line supervisor from the affected area or an adjacent work area. Workers may report unusual conditions or occurrences to their supervisor who, in turn, notifies upper management. An accident occurring in a public area, or on the perimeter of an industrial area, might first be reported by people passing through the area or by police and other emergency response personnel. However, once the initial notification is received, management must act promptly, decisively, and correctly.

Obviously, it is much easier for management to facilitate the notification process when accidents occur within the boundaries of company property, than it is for those that occur in public access areas such as highways or parking lots. To expedite the reporting process within work areas, accident reporting instructions can be posted prominently at or near telephones or intercom locations. Figure 4-1 shows an example of a placard which can be used to assist callers who are providing initial notification of an accident. By prompting the caller for the basic information indicated on the placard, the receiver of the notification will obtain some of the essential data concerning the accident event and, at the same time, immediately initiate appropriate response activities. Many organizations use stickers with emergency telephone numbers that can be placed directly on all company phones. Figure 4-2 shows a sample of one such sticker used in an industrial

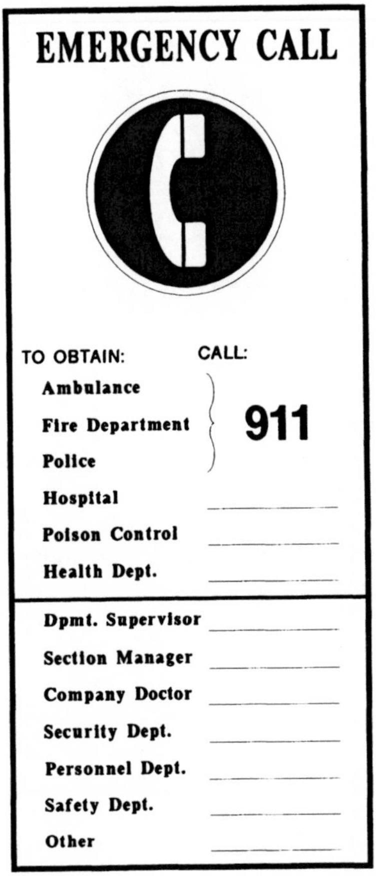

EMERGENCY CALL

TO OBTAIN:	CALL:
Ambulance	
Fire Department	911
Police	
Hospital	_____
Poison Control	_____
Health Dept.	_____
Dpmt. Supervisor	_____
Section Manager	_____
Company Doctor	_____
Security Dept.	_____
Personnel Dept.	_____
Safety Dept.	_____
Other	_____

helps to provide a uniform response to emergency persons

FIGURE 4-1.   An example of a placard used to assist callers who are providing initial notification of an accident.

facility. Placards and telephone stickers are a relatively inexpensive method of ensuring a standardized approach to accident/incident notification. Also, during the excitement that usually follows the occurrence of an accident, some people may not remember all the correct notification procedures and phone numbers. By providing the required information at or near a telephone, management can not only ensure that the accident will be reported to the proper authorities, but also it can facilitate its investigation process by immediately obtaining some essential information before it becomes distorted with the passing of time.

IN EVENT OF ACCIDENT OR EMERGENCY DIAL

911

Area Supervisor _____

Dept. Manager _____

Safety Dept. _____

FIGURE 4-2.   A sample of a phone sticker used to provide immediate access to important phone numbers in an industrial facility.

clarity
roles with coworkers!

Of course, immediate notification will be a useless exercise if those on the receiving end are not trained or aware of their responsibilities once the occurrence has been reported to them. This may sound obvious, but it is entirely possible that some people in the organization will not be aware of the necessary actions to take in the event of an accident. Some may not even be aware that their name or department appears on an emergency notification checklist. This has happened on many occasions. Therefore, it is extremely important that the telephone numbers listed on the placard or phone sticker be those of available people familiar with proper response actions. Company receptionists and switchboard operators should also be provided with emergency checklists with the names and telephone numbers of management personnel to be called in the event of an accident.

Newly hired employees should be provided instructions outlining proper accident notification procedures during their general orientation and training. However, it is a known fact that "new-hires" typically remember very little of the massive amounts of information provided during orientation. It is therefore recommended that follow-up or refresher training be provided on a periodic basis to ensure sufficient understanding of company rules pertaining to accident notification.

It is more difficult to ensure proper notification of accidents that *do not* occur on company-owned or operated property. After all, an organization's management does not have control over the actions of the general public. However, management can still take some basic steps to assist the process. Simply ensuring that the correct company name, address, and telephone number are prominently posted on all company equipment and vehicles routinely used for off-site tasks, could aid the bystander making an accident report. Here again, it is perhaps even more important to ensure that the numbers being called will access company personnel with the proper knowledge to respond appropriately.

Rescue Actions

The highest priority in any accident situation should always be the prompt rescue of endangered human life and the immediate treatment of those who are injured. Therefore, first responders at an accident scene should initially evaluate the nature and degree of any injuries. They should know how to determine what actions must be performed immediately to prevent loss of life, what can be done to relieve any further human suffering until injured personnel can be safely evacuated to medical facilities, and what steps can be taken to eliminate or reduce the danger of increased injury. First responders must have the training and knowledge to make such decisions under usually stressful circumstances. For example, unless personnel are in imminent danger or extreme distress, it is generally recommended that injured persons not be moved until professional response personnel have arrived. Minor injuries can be severely compounded, and major injuries can become crippling or even fatal, if improper movements are made. Personnel who arrive first at accident scenes where injuries have resulted must be aware of this possibility when assessing the nature and degree of those injuries. The recommended course of action is to provide any urgent and necessary care, such as a check for breathing, bleeding, or the need for cardiopulmonary resuscitation (CPR), to treat wounds using first-aid, treat for shock, and sustain life. All management personnel, who could be the first people of authority to arrive at an accident, as well as any other employee who could be a first responder, should therefore be trained and qualified in first-aid techniques (including CPR) to control the degree of loss during the immediate post-accident phase, such training being the initial step in the accident investigation and loss control process.

Aside from obvious moral and humane obligations, the prompt rescue of the injured and endangered is of critical importance to the overall loss control program. Response actions should exceed the reactive approach of providing treatment and move towards the proactive approach that attempts to prevent injuries. For instance, certain equipment, if left operating after an accident has occurred, could further endanger the injured as well as the safety of rescue workers. Responders must be capable of determining the correct course of action in such situations to ensure the minimum potential for additional loss resulting from the accident. This is particularly true with electrical equipment. If an explosive atmosphere exists as a result of an accident, the danger associated with shutting off the equipment (which may produce a spark) may outweigh that caused by allowing the equipment to continue operation. The potential for subsequent injury following the initial loss event requires immediate consideration and evaluation. Remember the domino theory discussed earlier. The threat of loss does not always end with the occurrence of the accident. It is entirely possible that additional dominos, downstream of the actual accident event, might be caused to fall as a result of actions taken (or not taken) following the accident event. When applicable, responders should survey the area for sources of explosive or toxic vapors, pressure release, radiation exposure, heat/fire, and other similar threats. It is possible that the force of energy exchanged during some sort of impact accident can have significantly loosened critical connections or ruptured transmission lines, allowing gasses to escape slowly into the atmosphere. As a general practice, smoking and the use of spark-producing devices should be strictly prohibited at all accident scenes.

All potential for secondary injuries in each work area should be evaluated in the planning phase. Proper considerations made during planning not only enhance rescue operations, they also help to ensure the preservation of evidence critical to the success of the investigation process.

Actions to Control Further Loss

As important as initiating actions to ensure quick and appropriate rescue of the injured is the control of additional, subsequent losses, such as injury to other personnel (including investigators), equipment, and property damage. Without proper consideration of control measures necessary to stabilize the accident site, property loss and injuries secondary to the accident event often far exceed that resulting from the original loss. The premature movement of injured persons or removal of equipment can result in a repeat accident or the compounding of injuries and loss. Furthermore, when materials or equipment similar to that involved in the accident are utilized in other tasks, it might be wise to consider suspension of their use until the investigation is complete.

Specific actions taken in response to an accident event to eliminate or reduce the potential for secondary loss will depend, of course, upon the nature of the accident. However, the following are examples of some measures that, if properly executed, can result in diminishing the likelihood of additional losses:

1. Maintenance personnel and equipment could be summoned to the accident site to assess the extent of any hazards resulting from damage to power lines, gas and fluid distribution systems, and the structural integrity of buildings and equipment.
2. Engineering personnel should be brought in to evaluate the need for any bracing of structures or other stabilizing measures and to determine the

appropriate engineering methods and equipment needed to accomplish these requirements.

3. Any special equipment needed to remove and repair damaged property, without causing further destruction in the process, should also be determined at this point.

4. Assessment of special response requirements must be determined as soon as possible. For example, the removal or safe rendering of explosives and other dangerous, toxic, or highly volatile materials should be completed by experts during the initial response to an accident event.

There is always the risk that a great deal of evidence could be lost or destroyed during initial response attempts to reduce the potential for secondary loss. Therefore, it is critical that those who respond to accident events, including technical experts, engineers, and maintenance personnel, be provided with an appropriate degree of training in the techniques associated with the protection and preservation of evidence. Even subtle actions, such as the activation of switches, valves, or other control mechanisms necessary to safeguard a situation, might actually be destroying or altering evidence crucial to the investigation of cause. Response personnel should first record in some acceptable manner (not memorization) the exact position of each control device at the moment they arrive at the scene. Responders should be capable of evaluating, not altering, the position of damaged equipment until it can be recorded by photograph, sketch, map, or some other more permanent method of written description. All work performed by any responder should be authorized and closely coordinated by the director of the investigation team. While it is best that individual investigators be significantly empowered to make the necessary decisions to conduct their assigned tasks, they should still be in close coordination with the investigation director. In order to ensure appropriate progress, reduce the likelihood of repetitive investigation actions, and guarantee the synergy required to complete a successful investigation, the director must always maintain an appropriate level of control over all actions performed by the individual investigation team members.

Initial Assessment of Available Evidence

As stated earlier, evidence can be lost to the investigation during the chaotic moments after an accident simply because it was removed from the site or significantly altered before anyone noticed it or made an accurate record of it. The actual moment the accident occurred (i.e., the *contact phase*) is usually a relatively brief moment in time. But the amount of activity generated following the loss event can be an extremely confusing and hectic period in which personnel *react* rather than *respond* to the situation at hand. Unless well-trained in accident response, human reactions can be somewhat irrational. For example, while immediate transportation of the injured for treatment may seem necessary for those seriously injured or in danger of further harm, it is not always recommended since injuries can be further compounded if the victim is moved the wrong way. Often, equipment and other items are quickly moved about or displaced in an effort to reach or treat injured personnel, create a passageway, or restore operational work. Such actions may alter the placement of physical evidence, the examination of which may be essential to the determination of accident cause.

In order to ensure the maximum preservation of critical evidence with minimum potential for secondary loss, managers at the scene are in the best position to exercise responsibility in identifying any key evidence before it is disturbed or changed. Simply by virtue of their ability to control others, managers on the scene become essential in controlling losses resulting from the accident. Their presence at and knowledge of the accident event also makes them invaluable to the investigation process.

A manager's initial identification of evidence should include, as a minimum, the following information:

1. A listing of the people involved (including the injured, witnesses, and those directly associated with the occurrence of the accident event).
2. Equipment involved in the accident (including equipment actually in use at the time the accident occurred and the operational status of each).
3. Materials involved (those being used when the accident occurred as well as any nearby that may have become involved due to proximity to the accident event).
4. Environmental factors influencing the accident (such as weather, illumination, temperature, noise, ventilation, etc.).

Accurate knowledge of these important factors may be crucial to the accident investigation. Unfortunately, these factors are also volatile, meaning that they are likely to change abruptly and dramatically following the accident, thereby making their initial identification by a responsible investigator even more essential. Typically, witnesses to an event cannot provide as accurate a record as the manager who has been trained to be observant of his or her surroundings. Therefore, the on-scene manager should ensure that, unless there is imminent risk of additional loss, nothing and nobody but the seriously injured should be moved, taken from, or added to the accident scene before accurate records detailing each aspect of the scene are completed.

Methods for Collection of Evidence

A primary function of the investigator is to locate and identify evidence relevant to the accident. Once available evidence has been collected, the investigator is to examine that evidence to determine its role during the accident event, reconstruct (if required) the exact sequence of events based upon the evidence collected, and develop viable conclusions from the evidence as to the cause or causes of the particular loss event. It therefore follows that sources of evidence after an accident, their immediate recognition and subsequent evaluation, are the most critical element of the entire investigation process.

Because the type of evidence will vary with each accident, not all methods of collection will apply to every accident situation. In each case, the amount and type of evidence available will require careful consideration. The process of evidence-collection begins almost immediately after the accident scene has been stabilized (i.e., the injured removed or treated and the threat of additional accidents reduced). In some instances, one investigator might be capable of collecting all evidence associated with the accident. Other more complicated cases, or those accidents which have generated a great deal of evidence, may require a team of investigators to perform the task of evidence-identification and collection. In any case, the investigator must exercise extreme care when attempting to identify, collect, retrieve, document, or otherwise handle critical pieces of physical evidence. Investigators not familiar with the fragility of some evidence might destroy it during the investigation process. Vital evidence could be lost or deformed due to poor handling, which will render that evidence useless, along with any secrets it may have concealed.

This concept of *fragility of evidence* has been understood for many years in the accident investigation and loss arena. Very simply, the term refers to evidence that can be easily broken, distorted, or lost, making it unusable or possibly even dangerous to use. Basically, evidence is considered fragile when the following conditions are possible:

1. Evidence is fragile in nature when it can be easily lost from view during an investigation, as items are misplaced, unintentionally covered up, or simply forgotten by eyewitnesses. Investigators must take special precautions to ensure no such losses occur.
2. Evidence is fragile when it can be easily distorted from its original shape or form, when it might be remembered incorrectly by witnesses, when it has been moved from its post-accident position, or when it has been willfully or inadvertently altered, bent or disfigured. All such evidence must be collected and protected as soon as possible.

3. Evidence susceptible to the recollection of more than one witness is considered fragile when one person's thoughts are conveyed to another. When people have enough time to rationalize what they have seen, the continuity of evidence obtained through witness testimony may be lost or broken. Also fragile is physical evidence such as a major mechanism when disassembled by an untrained investigator or another party wishing to *help,* evidence thus rendered useless to the investigation.

The investigator will typically discover fragile evidence in each of the categories shown in Figure 4-3. Sometimes referred to as the *four P's of fragility* in evidence-collection, *people, positions, parts,* and *papers* must be

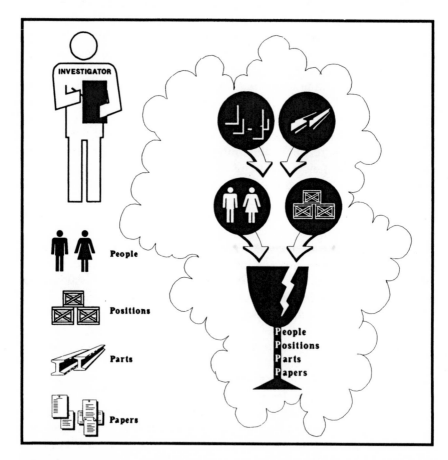

FIGURE 4-3.   The four sources of potential evidence (i.e., the *four-P's*) and the concept of fragility of evidence.

considered by the investigator in each of the three phases of the accident sequence.

People are the source of sensory (eye, ear, smell, touch) witness testimony to events that occurred in the pre-contact, contact, and post-contact phases of an accident (Bird, 1974). Their evidence, recorded primarily through individual observations and memories, is therefore the most fragile of evidence since its accuracy is subject to forgetfulness, rationalization, influence from others, and personal conflicts over what they may have seen. Even the manner in which they may have been interviewed can induce subtle distortions in the evidence.

Positions refers to the physical relationship, placement, and sequence of events as they occurred during the pre-contact, contact, and post-contact phases of an accident event. The post-contact position of evidence is particularly fragile since such evidence is typically moved or displaced by emergency response teams or others involved in the accident response. Positions are also subject to displacement by investigators, bystanders, clean-up crews, and production personnel wishing simply to return to normal work activities.

Parts can refer to wide variety of potential evidence. For example, there may be internal defects associated with a specific part that will not be discovered unless someone performs a detailed examination. Improper installation, violation of manufacturing standards or tolerances, the use of incorrect parts or materials for a specific item, and the willful or inadvertent abuse of parts are additional examples of parts fragility that could lead or contribute to an accident event. While certainly less fragile than the previous two P's, parts can be subjected to pilferage, corrosion, distortion, and misplacement.

Papers refers to any operating procedures, standard practices, job instructions, training records, maintenance records, inspection reports, and so on, background information that can also reveal evidence of what may have contributed to a loss event. It is the information recorded on paper that becomes evidence (i.e., the "paper trail"). The least fragile of the four P's, since they are somewhat difficult to change or alter, papers are often overlooked as sources of evidence, however. When such evidence exists, policy should require the impoundment of papers to prevent misplacement or alteration.

Security and the Preservation of Evidence

Thus far we have discussed the importance of collecting and preserving evidence exactly as it is discovered at the accident location. Preservation techniques include photography, sketches, maps, notes, witness statements, etc., all of which will be discussed in further detail throughout this text. Depending upon the nature of a particular accident event, preservation of evidence may also require additional actions to ensure its security and that of the scene prior to and during the investigation process.

Once the threat of a secondary accident has been removed or controlled, management must initiate whatever precautions deemed necessary to ensure the preservation of the accident scene. Company policy becomes the driving force in the preservation of evidence. Management must establish clear objectives that indicate the exact importance of post-accident activities with regard to the securing of evidence. For example, if an accident is given a low priority by management, it is unlikely that any serious effort to preserve the scene or the evidence will occur. Conversely, an accident considered high-profile due to public access, high dollar value, etc., may be secured for as long as the investigation team deems necessary. When an accident occurs on public or another party's private property, the task of preserving evidence can be further complicated. In many cases, certain legal negotiations will probably be necessary to ensure temporary access to the property during the investigation. Site security measures, although potentially more difficult to enforce, will be extremely important in such cases. The point is that management must provide the guidance (policy) needed to determine security requirements and the resources needed to accomplish these objectives.

In all cases, specific actions must be taken to preserve the scene as it was right after the accident occurred. All people, except investigators, should be instructed to vacate the area. Witnesses and those personnel involved in the accident should be removed to an isolated area, away from other personnel, the media, etc., to be interviewed as soon as possible. Depending on the accident circumstances, it may even be necessary to isolate witnesses from each other. Such measures help ensure totally unbiased statements, free from the rationalization that often occurs once people begin discussing what they have seen with other witnesses. Any other potential witnesses and bystanders should be identified, and their names and addresses taken, before they are allowed to leave the area. Employees or plant security personnel (if applicable) should be used to secure or even barricade the area to prevent curiosity-seekers from destroying essential evidence. In fact, to ensure that evidence is not moved, removed, or defaced, the scene should be totally controlled, such that no people, other than the investigators, will be permitted into the area without proper authorization from the investigation director. Even upper management must be prevented from disturbing evidence. It is not uncommon for the senior management of an organization to enter innocently and subsequently violate established control areas, unknowingly tinker with critical evidence, and move or alter the location of seemingly unessential items or materials, in an effort to determine for themselves exactly what happened. For these reasons, site preservation and security must be provided for in formal company policy, agreed upon by management, and planned properly before investigations become necessary.

In addition to ensuring the preservation of evidence through adequate security measures, other reasons for providing site security during an accident investigation might include the following:

1. To protect high-value or critical equipment involved in the accident.
2. To prevent adverse reaction from the public or employees to certain aspects of the accident.

For investigators, preservation of evidence begins with their initial walk-through inspection of the accident site. This event should occur as soon as practicable after the accident event. In a high-use, public access area, which may prove difficult to secure from crowds, the walk-through may be the sole opportunity for the investigator to preserve the location, position, and post-accident status of essential evidence at the scene. In such instances, or if items of evidence are likely to be moved (e.g., injured persons, items light in weight, or high-value property, etc.), the location of each should be recorded immediately or marked by the investigator during the walk-through.

PUBLIC RELATIONS

Many accident events, even those considered minor in nature, become the subject of great concern to the local community if they learn of its occurrence. In some cases, the accidents may have no impact on the public whatsoever. But when citizens become aware that an accident has occurred in a neighboring factory or other private facility, some will still insist on being informed of all the details. Much of this interest can be more than just idle curiosity. Events that adversely affect a company are potentially damaging to many other interested parties as well. For example, the local community may depend upon the company as a major contributor to its economic base. Suppliers and vendors also have interests associated with current and future business prospects within the facility. Customers are concerned with regard to price changes that may result from added costs created by the accident. Shareholders have an obvious interest, pertaining to the viability of their investments. Financial institutions have interests in the security of existing and future business loans to the company and personal loans to employees of that company. Competitors, labor unions, and certain government agencies are all interested in the events that affect a company's operations. With the advanced communication techniques and instant on-scene reporting capabilities available to the media today, it is entirely possible for all interested parties to become instantly aware of an accident event at a given company. It therefore becomes essential that organizations consider public relations during their investigation planning process.

Proper communication with the public is of particular importance in accident investigation, to ensure protection against premature and erroneous statements about the accident cause. All too often, overanxious reporters, wishing to "scoop" other reporters, will provide as fact their own interpretation of accident causes, based upon the limited information that may be available. Once such an announcement has been broadcast to the public, it will be difficult if not impossible to counteract the damage such erroneous information may cause. When a company allows for and plans proper public relations, the results demonstrate a good-will effort to ensure that the local community is properly informed of events which may or may not affect their lives. It will also show a level of compassionate concern for the people involved, as well as the relatives of those who may be injured.

The proper management of public relations, including the immediate release of news information, ultimate disclosure of accident cause, remedial actions and long-term safety management of the work system, is an essential aspect of accident investigation. To ensure control of accident losses, all aspects of public relations must be conducted in as disciplined and methodical a manner as the other aspects of the investigation process.

One thing is certain, companies that do not consider public relations an essential element of the accident investigation and loss control process will certainly suffer the detrimental effects that the uninformed media and the public can inflict following an accident event. There is nothing more potentially damaging to a company's image (and long-term success) than a perception of mistrust by the general public.

Informing Relatives of the Injured

News of an injury to a loved one can be a traumatic experience resulting in anything from a silent stare to an emotional collapse. It is extremely important that such news come directly from the company, not the news media, and preferably by personal contact from a company official specifically trained to handle such delicate situations. Notice directly from the company will ensure that the relative is provided factual and up-to-date information on a personal level. Whereas, initial notification of an injury or death of a loved one through a news report is rude, cruel, and inconsiderate of individual emotions. Personal contact with relatives allows for consideration of their reactions to the news. Face-to-face notification will also clearly demonstrate the company's concern for the injured, which time and again has helped improve human relations with all employees. Just the presence of a supervisor or designated company official is very meaningful to and appreciated by the relatives of the injured. This is especially important in cases of serious injury or death.

Companies can facilitate this notification process by ensuring that all newly hired employees indicate on their employment application the designation of the person to be notified in the event of an emergency. Of major importance throughout this process is the human relations element. It is essential that company officials approach the task of injury or death notification from a compassionate position of sincere concern. When possible, it is beneficial if the company seeks the assistance of a close friend of the victim or a clergyman to help inform family members. The designation of people to notify, as well as those who will assist, in the event of an emergency should be reviewed and updated periodically. An employee's annual performance appraisal provides an excellent opportunity to conduct this review. Employees should be encouraged by supervision to ensure this information is current. After an accident, unnecessary delays can occur trying to locate new addresses or trace family members. Such delays seriously complicate the notification process. Family members who feel as though they have been misinformed, or who receive initial notice through the news media, might seek legal action against a company to compensate for the emotional distress brought on by the accident.

Dealing with the News Media

An accident resulting in death, serious injury, or significant property damage is bound to attract the interest of the local news media. It is therefore in a company's best interest to initiate contact before the media becomes aware of the accident from some other source. This allows for control of information released by providing a central point of contact with whom the media may direct all inquiries. Prompt and willing disclosure of unfavorable (as well as promotional) news will serve to enhance overall relations with the media. Remember, in the wake of an accident, the issue of preventing further loss is of utmost importance. Alienating aggressive news agencies could result in serious long-term damage to a company's image, which in turn will have a detrimental effect on long-term business operations.

In most cases, the designated point of contact for accidental (or even a natural) death occurring on company property, or during company operations elsewhere, should be the normal public relations personnel. Therefore, pre-planning should include a policy on information disclosure. Any training that addresses critical and catastrophic emergency response should include public relations personnel. All managerial personnel should be trained in the proper channels of communication that have been established for disseminating information to the news media. There is nothing more damaging than an erroneous statement made by a quoted "company official" to the media without the authorization of a designated public relations person. All employees should be provided with guidance on the appropriate conduct to be exercised during media interviews at the scene of an accident. All too often, an employee's thoughtless speculation and conjecture over accident cause will become a factual report on the evening news.

A company should view the curious media as allies in accident response rather than the meddling, interfering, and prodding reporters that they often appear to be. Prompt and accurate disclosure of facts about an accident or death prevents the spread of rumors and speculation through the company and local community. When company personnel who are trained in dealing with the media prepare the news report, they can prevent errors often made by the media through inappropriate emphasis on a single event or aspect of the accident. Information provided by a hospital, coroner, police officer, other external agency, or even a bystander can also result in errors. Loss control includes controlling those potential losses that will result once such information is released to the public at large. Obviously, the best way to control these losses is to control the information provided to the media. Ill-chosen words can impart unintended connotations. When publicized through the media, these can prompt the unscrupulous to allege injury and compromise legal defense of the company's interests. There are thousands of law firms located throughout the country that pay close attention to information provided through the news agencies. Misinformation can lead to harassment of employees by the unethical seeking self-benefit through excessively high legal fees.

Regardless of the nature of the accident or degree of loss incurred no member of a company, management or otherwise, should ever make any statement regarding or even suggesting cause or fault. The primary purpose of investigation is to identify accident cause so that actions can be taken to eliminate the possibility of future, similar loss events. Premature conclusions to satisfy the news media can only compromise the interests of the investigation, the company, and the individual making the statements.

It is also essential that designated public relations officials be provided with up-to-date information regarding the accident event. While this sounds obvious, it is not uncommon to encounter situations where the investigation team may simply neglect to communicate prompt information to the one individual outside of their team who really must know: the public relations officer. It is extremely embarrassing for the company and the public relations officer when the media learns essential information from *other* sources, after first making inquiries to the designated company official. Investigators must remember to ensure that all appropriate company personnel are kept well-informed as the investigation process moves towards completion.

Release of Photographic Evidence

While photography can be essential to the investigative process, accident photographs should be carefully examined in detail by company safety, legal, and public relations personnel before allowing their use by the news media or in company reports that could eventually reach the media. This policy should include all photographs taken during all three stages of the accident (pre-contact, contact, and post-contact). Photographs that reveal a substandard condition or practice, such as a missing machine guard or the lack of personal protective equipment, could be used as evidence against the company during any subsequent litigation. For these reasons, photographic evidence should be treated as company-sensitive information with access strictly controlled.

Also, some companies may have conducted media tours of their factory or plant at one time in the past. The media usually archives any photographs or films they may have taken during such tours, of factories, equipment, award ceremonies, and so on. When an accident involves the equipment or people who may have been included in these previous photographic excursions, such photographs can become particularly strong evidence against a company. They are often used as evidence of general negligence in practice or attitude. Even when these photographs are unrelated to the specific accident, the media is skilled at creating imagery and lasting impressions. Chapter 7 provides a comprehensive discussion on photography and its use during the accident investigation.

Internal Release of Information

Caution should also be exercised when releasing accident-related information within the company. Such information should be made available as soon as possible to ensure that employees are informed of the essential elements of the accident event, but employers should exercise care over the methods they use to communicate accident facts internally. The status of the injured should also be promptly relayed to fellow employees, while ensuring the privacy of the injured is not violated in the process. To facilitate, control, and standardize the orderly release of information, accident reports such as that shown as Figure 4-4 should be used. As with all information pertaining to

```
████                                            ████
██  ████  ACCIDENT ALERT  ████  ██
████                                            ████
```

1. BASIC INFORMATION.

☐ PERSONNEL INJURY   ☐ PROPERTY DAMAGE OR LOSS

☐ PERSONNEL DEATH   ☐ OTHER INCIDENT

LOCATION: _____

DEPARTMENT: _____

DATE AND TIME OF OCCURRENCE: _____

2. NATURE AND EXTENT OF INJURY, ILLNESS, OR PROPERTY DAMAGE (INCLUDE COSTS, WHEN KNOWN).

3. PERSONNEL INFORMATION.

4. BRIEF DESCRIPTION OF INCIDENT.

5. INITIAL DETERMINATION OF APPARENT CAUSE(S).

6. AUTHENTICATION.

_____   _____
INVESTIGATING SUPERVISOR   DEPARTMENT MANAGER

DATE: _____   DATE: _____

FIGURE 4-4.   Sample Report Form for use in disseminating accident information in a timely fashion to other interested parties.

an accident, these forms should be reviewed by safety, legal, and public relations personnel prior to release. Prompt and accurate information provides employees with essential facts and helps to prevent the spread of damaging rumors. It also controls the amount of uninformed speculation that reduces work efficiency, assures employees that management is considerate of the workforce, and disseminates accident-prevention information at the same time (i.e., if other employees are made aware of the facts pertaining to the accident event, they are more likely to exercise additional caution when performing similar activities).

Regardless of the magnitude or degree of a particular accident event, the astute organization concerned with prolonged business operations in their respective communities, as well as the long-term preservation of employee morale, will consider the potentially adverse effects of poor public relations during and after an accident. Employers often utilize the media to promote events considered beneficial to their company's image and standing. However, they often neglect to provide these conduits of communication with vital and factual information pertaining to accidents. Indeed, proper public relations is a critical element in the overall control of losses resulting from an accident event of any degree.

WITNESS INTERVIEWS AND STATEMENTS

Earlier in this chapter, the concept of *fragility of evidence* was introduced, and evidence obtained as witness testimony from *people* was considered to be the most fragile. In fact, the most difficult type of evidence the investigator must obtain and preserve is that located in a person's memory. Even equipment that operates automatically, independent of human interference, was conceived and designed by humans and is therefore not immune to the "people" category of evidence collection. However, since almost every system, activity, operation, task, or function has a human element, the evidence obtained here is of vital importance in determining the basic causes of an accident. *People* evidence, the first of the *four P's of fragility* in evidence collection, is obtained through witness interviews, statements and examinations.

A *witness* can be defined simply as any person who has information relating to the accident. This includes anyone from those persons principally involved in the accident (i.e., those who were injured or those directing or controlling the task that resulted in the accident) to those who may have seen or heard the accident or observed the work environment at the time the accident occurred. A witness may also be someone who had knowledge of the events occurring during any of the three contact stages of the accident. Therefore, in order to ensure the proper identification and interview of all potential witnesses, the investigator must thoroughly examine every aspect of the work environment where the accident event occurred. For example, witness testimony from someone in an adjacent area, who may have heard unfamiliar sounds coming from a machine just prior to its failure, cannot be overlooked. Depending on the nature of the accident, such testimony may be the only witness evidence available to the investigator.

Understanding Human Behavior and Learning

For investigators to adequately identify witnesses, interview each, and evaluate the evidence obtained through witness statements, they must first understand some fundamentals of human behavior. This includes a familiarization with the process of learning through observation. Basically, people differ from one another in their individual abilities to recall specific details of a given event. This difference is based primarily in their powers of observation and seldom has anything to do with their degree of intelligence. Individual intelligence may affect the ability of a particular witness to adequately articulate what he or she may have seen and heard, but it does not substantially influence the facts contained in their testimony.

Numerous studies in the psychology of human learning have shown that people—regardless of who they are, how old, what sex, degree of intelligence, and so on—experience true learning as a result of six basic categories of information exchange. These categories, also known as the *laws of learning*, each dramatically affect the human traits of observation and memorization. Few people are trained observers. Most people do not really *see* what they are looking at and rarely remember the details of what they have seen with a high degree of accuracy. Therefore, the investigator must carefully evaluate the value of witness testimony. If the task of collecting witness evidence is approached with a clear understanding of how the six laws of learning affect memory, then it is highly probable that the information obtained will be much more useful to the overall investigation. Figure 4-5 shows the six laws of learning as *readiness, exercise, effect, primacy, intensity,* and *recency.* Each has a different effect on the amount and clarity of a person's observation and subsequent memorization of events, as follows:

1. *Readiness* implies that people must first be ready to learn and remember what they have seen. This often occurs routinely in a work setting where alert people are performing job functions requiring concentration and a genuine openness to the information about their assigned work. Complacency and boredom, however, are distractions that will substantially

reduce this element of the learning process and subsequently adversely impact the remaining laws.

2. *Exercise* means that there must be some repetition of facts or information-exchange in order for a person to recall what he or she has seen or heard. This law is used primarily in a classroom or some other formal educational setting and does not often come into play during an accident event. It is also encountered by most everyone who has listened to a particular song over and over again; before too long, we know all or most of the lyrics without ever having read it. The investigator should be aware that a witness who mentally recites the facts of his or her testimony over and over again in preparation for an interview, is actually exercising his or her memory. If too much time passes between accident and interview, people will often begin to interject their own conclusions and conjecture during this process of mental exercise. As a result, distortion of the true facts can easily occur.

3. *Effect* means that the particular observation must have special meaning or subjective significance to the individual. Very simply, people will best remember those events that have the most meaning to them. Sometimes a religious event, a travel experience, or simply a book can have such an

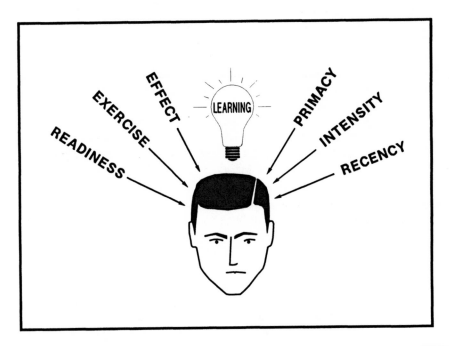

FIGURE 4-5.   The six laws of learning.

effect on people, solely due to their values in life, that they commit this information to memory when others may not. This aspect of the learning process is probably the most unpredictable since everyone has a different set of values. What is important to one person may mean absolutely nothing to another. Investigators must be considerate of this individuality when interviewing a witness, especially when testimony between one witness and another differs.

4. *Primacy* can be one of the most useful laws of the learning process in terms of accident investigation. It implies that people will often recall quite readily those events which occurred first in a chain of events. In fact, when this type of learning occurs, subsequent events do not typically have a great deal of influence on memory and the person may not remember them as quickly. Investigators should explore testimony to accurately determine the facts if it is suspected that primacy in learning has occurred. Usually, the events recalled as occurring first will most likely be the most accurate.

5. *Intensity* results from an extremely vivid experience. Except for the pre-contact stage, most accidents will fit this category of learning. This law revolves around the premise that the more intense an experience, the more detailed the recollection. Vietnam War and other war veterans sometimes suffer from what has become known as post-traumatic stress disorder (PTSD). Events so vivid and intense in nature and degree have influenced the memories of these veterans that they recall from memory images so real and disturbing that they become physically violent or ill. Particularly gruesome or destructive accidents can affect a witness in a similar way, and investigators should therefore use caution when asking people to describe or recall the details of such events.

6. *Recency* is another quite useful learning law in terms of accident investigation. It states that people will best remember the last or most recent facts or events. This is the primary reason that an observer's testimony is considered most credible and factual immediately following an accident event. Remember the law of *exercise* discussed earlier. If people are allowed to rehearse their testimony in their own minds or with other witnesses, distortion of actual facts can occur due to the subconscious rationalization and speculation that often results during this rehearsal process.

In most accident scenarios, those persons involved or present seldom observe *all* that has transpired. For instance, the events that typically occur during the pre-contact stage are usually routine in nature and happen without much notice from those working in the area. Witnesses to the pre-contact stage may not feel that their testimony is even worth mentioning since they attach no importance to the seemingly routine events that occurred before the accident. After all, if events during the pre-contact stage had been so different from the routine, people might have noticed and taken appropriate actions that might have prevented an accident from ever occurring. Whereas, during the aftermath of an accident, people are not prone to record all the activities that occurred in the often-chaotic moments of the post-contact stage. Investigators should therefore learn to phrase questioning in such a way as to stimulate recollection and inspire people to think about *all* of what they have seen and heard, not dismissing any facts, even those that a witness may have already concluded to be inconsequential.

Locating Witnesses

Since the value of most witness statements is highest when derived from testimony gathered immediately after the accident occurs (e.g., the laws of recency and intensity), it is essential that interviews begin as soon as possible. However, before any interview can take place, the investigator must first locate those individuals who have information pertaining to the accident event. Prompt arrival at an accident scene can certainly facilitate the investigator's task of identifying those in the immediate and general area who might be potential witnesses. However, in some cases the investigator may have to do some investigating just to locate witnesses. Some suggestions for seeking out people who may know something about the accident include, but are not limited to:

1. Those directly involved can often name those who were there at the time the accident occurred.
2. Known witnesses (i.e., those obvious or already identified as people who saw or heard what happened) may be able to identify others who were in the area. An interviewer's questions should include inquiries about other potential witnesses. A nervous witness may neglect or forget to mention these other people if not directly asked. Also, most organizations require work schedules, area sign-in logs, or time cards. These are excellent sources of information when trying to determine who may have been present at the time of the accident.
3. Area supervision is often there during an accident, or first at the scene after it has occurred. Such area supervisors, then, are an excellent source of information pertaining to the identity of those who may have witnessed the event.
4. People working in adjacent areas may have information on sights, sounds, weather, lighting, noise, and other environmental factors that may have changed immediately after the accident.

5. Emergency response crews are trained observers, often recording names of those who were present when they arrived.

Depending on the particular accident situation, an investigator can also research other available non-company records to identify witnesses, such as:

1. Police record witnesses to traffic accidents. Researching their notes can help identify others who were in the area when such an accident occurred.
2. Hospital records might show those who left the scene early to obtain treatment or others who showed up later for minor injuries. Anyone who was injured during the accident should have some level of information to share with an investigator.
3. News reporters may have located witnesses to support their coverage of the accident. Sometimes, these witnesses do not appear in the final news broadcast, but their identity will often have been recorded by the reporter during the interview.

Most of these sources are immediately available to assist the investigator in identifying and locating witnesses. Others may require more time and effort. Whatever the case, the investigator must know when to stop looking for witnesses and when to start interviewing. Remember, if too much time is allowed to pass after the accident has occurred, the investigator runs the risk of loosing credibility of witness testimony. Ideally, there should be more than one person working to identify and interview witnesses, one investigator beginning to interview right away while another continues to locate and identify witnesses.

Ensuring Unbiased Testimony

It is rare for any one witness to a given event to be able to recall clearly *all* the details of what occurred before, during, and after that event. Realistically, each witness will likely have a slightly different description of what he or she observed, as compared to the next witness. Their individual observations will differ for many reasons, such as:

1. No two individuals see details or objects and remember them the same way.
2. Points-of-observation vary from witness to witness.
3. Witnesses are different in both technical and personal backgrounds.

4. Witnesses are interested in self-preservation and the protection of friends.
5. Witnesses differ in their abilities to rationalize what they have seen and articulate an explanation or description.

Differing accounts of the same event may be provided by operators, technicians, supervisors, managers, and spectators. Consequently, it is not uncommon to find that initial testimonies from various witnesses will be somewhat disjointed. In the worst case, it may even appear at first as though each person witnessed an entirely different event.

For these reasons, it is strongly recommended that witnesses be kept from each other until after the initial interviews have taken place. If allowed to discuss the accident among themselves before the interview occurs, witnesses might recognize that there are information gaps in their accounts of the event. To avoid appearing as though one is not as good an observer as the next person, a witness may actually fabricate some of the details in his or her statements. The witness may also pick up details from anther person's account and adapt them to his or her story. This is especially true when another witness may be considered more technically competent. Although such reactions will allow a more clear and descriptive account of the event, it is clearly counterproductive to the investigation process. However, if a witness is kept from others until after they tell their stories, the investigator is more likely to obtain a more usable account of the events surrounding the accident. Remember, it is the investigator's responsibility to locate the missing parts of the puzzle.

Another consideration when conducting an interview with a witness is that of *self-preservation*. Some people have a tendency to alter the facts when they have the perception that their statements may threaten their own position (or that of a close friend), their continued employment, or their influence and acceptance among their peers. These factors all threaten the accuracy of the witness interview and, subsequently, the investigation itself.

Selecting the Interview Location

It is important that the interview of a witness occur in a comfortable area or atmosphere conducive to ensuring an accurate account of the accident event. An interview at the accident scene has several advantages and should be attempted if the nature and degree of damage is such that an interview could take place without jeopardizing personnel safety. Some benefits to on-scene interviews include the probability of more accurate recall from witnesses who attempt to describe the details of the event from the location where they actually observed the accident as it transpired. As long as privacy can be maintained and there are no interfering factors such as noise, poor lighting, or excessive activity, the investigator should attempt to conduct the witness interview at the scene. If, however, the activity and environment at the accident scene cannot be controlled, the investigator must select an alternate location where the interview of a witness can be properly conducted.

When selecting a suitable place to conduct interviews, the investigator must ensure that the location is non-threatening to the witness. When a witness is psychologically uncomfortable about the location of the interview, the accuracy or completeness of his or her testimony might be affected. For example, an interview conducted in a high-level manager's or executive's office may intimidate the witness, especially if he/she is a lower-level employee not used to being in such a place. Even if the owner of the office is not present during the interview, there are still distractions caused by the psychological effect such a place has on some people. Even the supervisor's office cannot be considered neutral ground—being interviewed across the desk of an immediate boss is intimidating. Remember, the whole purpose of the interview is to obtain information about an accident from those who were there. It is therefore extremely important that the first time witnesses actually describe what they have seen, they should be as comfortable as possible with the interview, their surroundings, and their role in the investigation process.

A small conference room, break area (when not in use by other employees), company library, or reception waiting area are all possible locations for interviews. When a *first-time* statement is taken from a witness, it should be in a one-on-one (witness-interviewer) situation. Historically, such an arrangement has shown itself to be the most effective, an interview between two people in privacy. Group interviews can be intimidating to some witnesses. Of course, depending on the nature of the accident, the disposition of the witness, and the position of the company with regard to liabilities, one-on-one interviews may not be possible. A witness (or the company) may wish to have legal counsel present during any interview. Also, depending upon the nature of the accident, the various members of an investigation team may have specific and pointed questions to ask a witness. In this case, a group interview might be appropriate at some point after the initial interview has occurred, but caution in approach should still be exercised so as not to intimidate the witness into silence.

Beginning the Interview

It is important that the interviewer establish a friendly atmosphere with the witness at the beginning of the interview. The witness should be put at ease, made comfortable, offered coffee or refreshments, called by his or her first name, greeted with a handshake, and so on. Incidentally, the handshake can help the investigator discover something of the mood of the witness: a moist palm may sometimes indicate nervousness. Calluses reveal a person used to working with his hands. A limp grasp can show a timid, disinterested, or insincere person. A firm, strong grip from people who make solid eye-contact may reveal persons sure of themselves and ready to cooperate. The witness who quickly grasps a hand and attempts to squeeze the blood right out of one's fingers may be an aggressive personality. Of course, there are other explanations for the types of indications and behaviors discussed here. The astute investigator would never form conclusions based solely on a handshake. These are simply observations that may help the investigator form an opinion of the witness as the interview proceeds.

Proper communication between witness and investigator is the key to understanding. The investigator who cannot establish this at the beginning of the interview will fail. In order to succeed, communication must occur on the level of the witness, not the interviewer. It is essential that the interviewer speak in terms familiar to the witness. An accident investigation interview is no time to impress people with articulation and flowery phrases that are not common for normal conversation or are basically unfamiliar to the witness. From the onset of the interview, the witness must feel that he/she is being treated as an adult with something meaningful to offer. Interviewers must never speak down to a witness or conduct themselves in a manner that can be perceived as arrogant or even superficial.

After the witness arrives at the designated interview location, greetings have been exchanged, and the witness is made comfortable, the investigator should initiate the interview process with simple introductions and an explanation of the purpose of the interview. If the interviewer elects to record the testimony of a witness on audio or video tape, he or she must inform the witness prior to the start of the interview that such recordings are being made. However, it is not generally advisable to tape or video a witness statement since it can be extremely intimidating to most people. Even if the purpose is to ensure an accurate account of the testimony and even if a witness has granted his consent to use the device, the presence of a recorder makes some people very uncomfortable and quite nervous. Instead, the interviewer should be capable of taking extensive, abbreviated notes during the testimony, for transcription after the conclusion of the interview. Note-taking should be inobtrusive, done in front of witnesses but not distracting to them—but an interviewer should never attempt to make notes covertly or otherwise concealed from the witness. Discovery will create an atmosphere of distrust and a subsequent lack of cooperation.

As a matter of practice, the following tips are provided as suggested methods of initiating an interview. The investigator should:

1. explain who he/she is (if the witness and interviewer do not know each other already);
2. explain why the accident is being investigated;
3. discuss the purpose of the investigation (i.e., to identify problems and not to determine fault or blame);
4. provide assurances that the witness is not in any danger of being compromised for testifying about the accident;
5. inform the witness of who will receive a report of the investigation results;
6. ensure witnesses that they will be given the opportunity to review their statements before they are finalized;
7. absolutely guarantee the privacy of the witness during the interview;
8. remain objective—do not judge, argue, or refute the testimony of the witness.

Once the investigator has covered the above listed elements, the interview should be allowed to proceed without interruption. Initially, the interviewer should obtain routine information such as the name of the witness, his or her job title or function, experience and training level, number of years with the company (if applicable), and so on. Starting the interview with such basic and routine questions serves two purposes: 1) the witness becomes comfortable and, therefore, more confident in his or her ability to answer questions, and 2) the interviewer observes how a witness handles questioning and determines a person's ability to articulate a response to increasingly more difficult questioning. This subtle exchange should assist the interviewer in identifying the level of communication at which to conduct the remainder of the interview.

Once the routine questions have been asked and answered, the *real* interview should begin with a simple request for the witness to describe what they know about the accident event. Ask him or her to explain specifically what was seen, heard, and felt prior to and at the time of the accident. Find out what the witness did immediately after the accident occurred. Throughout the testimony, the witness should be allowed to describe their observations in their own manner. The interviewer should avoid any line of questioning that causes the witness to structure a response in some type of order. This hinders the spontaneity of the testimony and may actually cause a witness, more concerned over the proper structure of response, to exclude critical information from statements.

The interviewer must be an exceptional listener during the witness testimony. Much of the information obtained from a witness, while interesting, may not be usable in terms of investigating accident cause. Other aspects of the interview may be critical to the investigation if they are further expanded and discussed.

Expanding the Focus of the Interview

Once witnesses appear to have exhausted their recollection of the event, the interviewer should move directly into those areas where further development or expansion of certain statements of observations is required. All questions during this phase of the interview should focus on the *who, what, when, where, and how* aspects of the accident event. Such focus will preclude the possibility that the witness will provide opinions rather than fact. While the *why* questions can be asked at the end of the interview, the majority of the testimony should be based upon facts and observations and not on opinions. It is the investigator's responsibility to ensure that the interview remains on this track. When opinions are solicited, such as at the end of the interview, the investigator must still keep an objective attitude and avoid the desire to draw premature conclusions based upon the personal opinion of a particular witness.

As a minimum, the following areas should be covered during the interview of a witness:

1. *What* was the exact or approximate time of the accident?
2. *What* was the condition of the working environment at the time of the accident; before the accident; after the accident (e.g., weather, noise, temperature, distractions, etc.)?
3. *Where* were people, equipment, and materials located when the accident occurred; and *what* was their position before and after the event?
4. *Who* are the other witnesses (if applicable) and *what* is their job function?
5. *What* was moved from the scene, repositioned, or changed after the accident occurred (including injured persons)?
6. *When* did the witness first become aware of the accident and *how* did he or she become aware?
7. *How* did the response or emergency personnel perform, including supervisors and outside emergency teams?
8. *How* could the accident have been avoided?

Questions 1-5 deal with the *facts* of the accident. This information is required to properly complete the accident report as well as validate the credibility of the witness testimony. Generally, the investigator will already have some knowledge of the time, environment, position of equipment, other witnesses, and so on. If the witness provides testimony that significantly differs from previously established factual information, then the value of his or her testimony should be questioned.

Question 6 will help the interviewer determine at what point during the accident occurrence the witness became a witness. If the witness indicates that he or she became aware that something was wrong because of a strange sound or unusual visual event *prior* to the actual accident, the investigator may have uncovered vital information about the cause of the accident.

Even though question 7 may solicit an opinion, witness observations about response teams can provide valuable feedback on the effectiveness of the organization's post-contact loss control program. The interviewer should carefully weigh information obtained during this line of questioning against established company procedures for emergency response. The response of supervision and management personnel to accident events may also indicate a need for additional training and emergency response drills to enhance the overall loss control program.

Question 8 is a deliberate attempt to have the witness volunteer an opinion on the cause of the accident. While his or her statements will by no means conclude an investigation, such questioning will draw upon the individual's experience and background to determine viable solutions to the problems the witness perceives as the cause of the loss event. A response to this question should be used as a point of reference at the end of the investigation. An opinion about accident prevention may actually provide a direction from which the investigator can proceed when providing recommended solutions and corrective actions.

Concluding the Interview

Once all questions have been answered and the witness appears to have nothing left to offer, the interviewer might close the session by asking the witness to provide some suggestions on accident prevention. This line of questioning is best left to the end of the interview since it deliberately requires the witness to draw conclusions, introduce inferences, and provide opinions. At this point, the interview is no longer a fact-finding activity. But, allowing the witness to elaborate opinion at the end of the interview will help foster an attitude of accident prevention. Remembering the law of learning known as *recency:* the witness might well remember his/her own thoughts regarding accident prevention if they are one of the last things discussed. The interviewer may also obtain additional information from witness opinions that indicate the necessity for improvements in the areas of training, emergency response management, or any other aspect of the organization's infrastructure.

When witnesses offer no suggestions, the interviewer may either prompt them again for their opinions or simply thank them for their assistance and conclude the interview. Witnesses should once again be reminded that their assistance is invaluable to the investigation. They should also know that, at any time after the interview, they are more than welcome to return if they should think of something in addition to what they have already provided in testimony. They should be made aware that their assistance in resolving the problems which led to an accident is of great help to the company, their fellow employees, and themselves.

After the Interview

When all witnesses have been interviewed and their testimonies documented in written summary, the investigator should allow witnesses to review their statements to ensure that the written description of what they said is, in fact, what they meant. Very few people are expert stenographers, including most would-be accident investigators. Allowing the witness to review what has been written by the investigator will serve a dual purpose: It will verify the accuracy of the recorded testimony, and it will also allow the witness to further explain and clarify certain aspects or facts about the accident they may have overlooked during the actual interview.

Before any witness testimony is used to help draw conclusions about the cause of an accident, the investigator should carefully isolate and evaluate the specific *points of fact* contained in the witness statement regarding their exact applicability to the investigation. Of all the things a witness may discuss, insinuate, or otherwise mention during an interview, there may be very few factual statements that will ultimately aid the investigation process. While much of what he or she has to offer may be extremely interesting and important, especially in the mind of the witness, his or her statements may not provide very useful or new information. More likely, much of the testimony may serve only to confirm what is already known about the event. The investigator must therefore be adept at quickly and accurately sifting through the various pieces of information to uncover only those facts that may provide the most benefit in determining the circumstances surrounding the loss event. Once those facts have been identified and isolated, they can often be verified by comparison with the physical evidence at the scene. For example, if a witness claims that a light-weight object was moving relatively slow over a great distance before it collided with another stationary object, then evidence of minor collision damage at the accident scene would substantiate such testimony. However, extensive damage would be more indicative of a heavy object, moving rather quickly before its collision with a second object. The point is, when the physical evidence does not corroborate the stated evidence, the investigator must further analyze the discrepancies between the two in order to fully understand the true value of each to the investigation process. On the one hand, alteration of the physical position of evidence, attempts to change the degree of damage, or the repositioning of switches or controls may have occurred after the accident in an effort to mask unsafe conditions. On the other hand, a witness may not have been totally sure of the exact circumstances about the accident and simply *created* certain details to fill in perceived gaps in his or her testimony. Whatever the reasons for discrepancies between stated evidence (i.e., witness testimony) and physical evidence at the scene, the investigator must be able to recognize and evaluate each to determine the necessity for further investigation (i.e., follow-up interviews).

Follow-up Interviews

One method to further examine points of conflict between various testimonies and physical evidence is to conduct a second interview. If this is done, the interviewer must ensure there is no implication that the witness is considered to have been untruthful, concealed facts, or made mistakes during the first interview. Instead, the witness should know from the beginning that the second interview is necessary to clarify key points of fact so that the investigation does not proceed in the wrong direction. Attempts to coerce or lead the witness must be avoided. The witness need only explain once again the details of what he/she knows of the accident. The interviewer must be aware that, with the passage of time, people have a tendency to *rationalize* the events they have witnessed. As a basic characteristic of human behavior, rationalization allows us to further understand (or *make sense* out of) certain unfamiliar, frightening, or simply irregular events we have experienced or witnessed. As we consider these events over and over again, we relate them to things we already understand. This *understanding* may be based upon past experiences, personal values, preconceived assumptions, or individual opinions. As we attempt to further this understanding, we eventually begin to draw conclusions that may or may not have any basis in actual fact. The more time that passes between the occurrence of the event and the interview, the more likely it is that a person has had the chance to substantially rationalize the situation and circumstances surrounding that event. Without knowing it, a witness may begin to incorporate these opinions and conclusions (arrived at through rationalization) as fact during his or her testimony. The more they recount or repeat their statements, the more solid this non-factual evidence becomes to the witness. The interviewer must therefore be cognizant of this possibility, especially during any follow-up interviews that might be performed much after the accident event. Also, legal proceedings subsequent to an accident event usually occur much later, sometimes years after the occurrence of the accident. Witnesses called to testify in such cases often have great difficulty remembering the specific details of the accident or the statements they may have made during the initial investigation interview. Rationalization, coupled with the added stress of a courtroom situation, can reduce the value of a witness testimony to near worthless.

To help a witness properly recollect the event and, hopefully, preclude the possibility of rationalization, it is often suggested that the interviewer have the witness complete a written statement immediately *after* the initial interview. Then, if a second interview or litigation is required at some time in the future, the witness can refresh their memory of what was said originally by reviewing their own written descriptions. Having witnesses provide a written statement at the completion of the first interview, rather than at the beginning, will cause them to think about what has already been said. Since oral statements made during the actual interview are usually spontaneous answers to previously unknown questions, the witness has had little time to rationalize or otherwise prepare their statements. Asking them to write a statement after the oral interview will therefore afford them the opportunity to further solidify what they stated orally. However, if they prepare the statement first, they will more than likely concentrate on the structure of their statements. Then, when the oral interview occurs, they may be more concerned with following their written statements than they are with the actual answers to the questions being asked. Spontaneity is lost and the value of the testimony is decreased.

Figure 4-6 shows a sample Witness Statement Form that can be used to collect and record various witness statements. Use of such forms helps to standardize the process as well as provide structure for the witness providing the statement. A word of caution about the use of written statements: In some instances, a witness may refuse to write and sign anything pertaining to the accident. They may perceive the use of a written statement as a threat to their own employment security or may wish to avoid getting more seriously involved in the situation, especially when there is a likelihood of litigation. If this occurs, the interviewer must be careful not to further alienate the witness. Immediate attempts should be made to reconcile the situation and obtain whatever statements the witness may still wish to provide. In such cases, the interviewer must rely solely upon their own written notes taken during the interview as a record of the testimony.

When a written statement is to be used, the interviewer should explain the intended purpose and use of the statement and, more importantly, who

WITNESS STATEMENT

SECTION 1: GENERAL INFORMATION	SECTION 2: WITNESS INFORMATION
Date of Interview: _____	Witness Name: _____
Time of Interview: _____	Address: _____
Duration of Interview: _____	_____
Location of Interview: _____	Telephone: _____ Age: ___ Sex: ___
Name of Interviewer: _____	Social Security Number: _____
Title of Interviewer: _____	Employer: _____
Date of Accident/Incident: _____	Address: _____
Location: _____	_____
_____	Employer Telephone: _____
Accident/Incident File Number: _____	Job Title or Function: _____
	_____

SECTION 3:   WITNESS STATEMENT

I, _____ , do understand that the primary purpose of this statement is in the interest of accident prevention. However, I understand that the information provided by me below may also be required to be revealed in a court of law. The facts and observations below are true to the best of my knowledge and belief.

_____
Signature of Witness

Signature of Interviewer: _____   Page____ of ____

FIGURE 4-6.   Sample Witness Statement Form.

will have access to it. This includes the possibility that the statement may be used in court. Witnesses should also know that they have the right to deny providing such a statement if they feel uncomfortable doing so.

The investigator should know that nearly half of all the information to be uncovered during an investigation will result from witness interviews. Obtaining true and useful information is extremely vital to the success of the overall investigation. The information obtained through witness statements becomes an important element of the final accident investigation report. With few exceptions, the witness interview is often a key to accident cause determination. Proper training and awareness on the part of the investigator is therefore essential.

THE ACCIDENT INVESTIGATION REPORT

All information gathered during the course of the investigation should be properly and formally recorded in an *accident investigation report*. Formal documentation serves both the loss control program as well as the organization's overall safety and health efforts. Because it provides a condensed version of the relevant facts, findings, shortcomings and recommendations in one package, the report affords each level of the management structure the opportunity to determine the type and priority of corrective actions and measures required to reduce loss potential. The report will also allow the safety and health professional to determine the quality of the organization's loss control program and assess management's ability to accurately investigate for accident cause. Documented accident investigation reports may also be the subject of review by federal or state safety agencies, during routine or specific inspections, to help them assess the company's safety programs. Also, unless specific actions are taken with regard to document protection under attorney/client privilege provisions, accident reports will more than likely be the subject of subpoena during litigation proceedings. For these and many other less obvious reasons, the accident report should be clear, concise, accurate, and complete.

Although there are no established or industry-standard formats for recording the accident investigation into a report form, there are several aspects of the process which are common to most reports. Each company will generally develop its own format for the investigation report, one that is acceptable to management. Therefore, the suggestions discussed here are intended as a guide to provide the reader with the basic information necessary for building an appropriate accident investigation report. In practice, the reader may wish to include additional information not discussed here as deemed necessary by each particular circumstance. Whatever the case, it is worth repeating here that any and all information recorded in the report should be clear, concise, complete, and, above all, defensible.

The Accident Report Form

Since documentation of the accident investigation actually begins with the filing of an accident report form, it is generally considered the foundation of the investigation report. In most cases, the initial accident report form (sometimes called the accident investigation report form) is filed by the responsible area supervisor or manager. This report then becomes the framework upon which any subsequent investigation activities will be developed. A well-designed form will establish a standard method for structured and consistent reporting within the organization. The information obtained on the form will answer all the basic but essential questions about the accident event such as who, what, when, where, and how. It will also define those areas which require further elaboration and investigation. A good accident report form will be divided into two primary areas: *general information* and *specific information* General information provides an assortment of facts that can be categorized and evaluated for trend or other types of comparative analyses. The specific information section will document information unique to the particular accident event (such as the work environment, people and equipment involved, and the procedures in use at the time of the accident).

Figures 4-7 (side A) and 4-8 (side B) are samples of an Accident Report Form to be completed by responsible supervision. To emphasize this responsibility, some organizations label the form Supervisor's Report of Accident or some similar title. Whatever the form used, and by whatever name, the common elements contained on this sample should be considered when developing accident report forms. This form, which should be completed by the responsible supervision as soon as possible after the accident event, begins the investigation process. If the investigation is initiated while memories are still fresh, equipment not yet moved, the work environment unchanged to any great extent, and the people involved still in or near the area, then the investigation has a greater chance for success.

After the information requested by this form has been obtained, the investigation can proceed into the detailed aspects of the accident event. The form will help identify those areas of the investigation that require additional evaluation using more definitive methods and techniques. These include the witness interviews, as previously discussed in this chapter; the development of maps, sketches, and drawings (as discussed in Part II, Chapter 9 of this text); photographs (Chapter 7); and any other activities (such as system safety analyses, engineering studies, metallurgical analysis,

ACCIDENT REPORT FORM

SIDE A

Capital Automation Inc.

CAPITAL AUTOMATION INCORPORATED
7786 SEMICONDUCTOR BLVD.
ANYTOWN, TEXAS 12345
800-555-1212

Case or File Number

Date This Report

GENERAL INFORMATION

1. NAME OF INJURED	2. SOCIAL SECURITY NUMBER	3. SEX ☐ M ☐ F	4. AGE	5. DATE OF ACCIDENT

6. HOME ADDRESS	7. EMPLOYEE'S USUAL OCCUPATION	8. OCCUPATION AT TIME OF ACCIDENT

9. EMPLOYMENT CATEGORY
☐ Regular, full-time   ☐ Temporary
☐ Regular, part-time   ☐ Nonemployee

10. LENGTH OF EMPLOYMENT
☐ Less than 1 yr.   ☐ 4 – 7 years
☐ 1 – 3 years   ☐ More than 8 yrs.

11. TIME IN OCC. AT TIME OF ACCIDENT
☐ Less than 1 yr.   ☐ 4 – 7 years
☐ 1 – 3 years   ☐ More than 8 yrs.

12. NATURE OF INJURY AND PART OF BODY INJURRED

13. TIME OF INJURY
A.M. _____ P.M.
☐ Time within shift?
Type of shift _____

14. SEVERITY OF INJURY
☐ FATALITY
☐ Lost workdays
☐ Restricted workdays
☐ Medical treatment
☐ First-aid, return to work
☐ Other, specify _____

15. NAME AND ADDRESS OF TREATING PHYSICIAN

16. NAME AND ADDRESS OF HOSPITAL (if applicable)

17. SPECIFIC LOCATION OF ACCIDENT

ON COMPANY PREMISES? ☐ YES ☐ NO

18. PHASE OF WORKDAY AT TIME OF ACCIDENT
☐ During rest period   ☐ Entering or leaving plant
☐ During meal period   ☐ Performing work duties
☐ Working overtime   ☐ Other _____

SPECIFIC INFORMATION

19. WITNESS INFORMATION

NAME: _____   NAME: _____   NAME: _____
TITLE: _____   TITLE: _____   TITLE: _____
PHONE: _____   PHONE: _____   PHONE: _____

20. TASK AND ACTIVITY AT TIME OF ACCIDENT

A. General type of task: _____

B. Specific activity: _____

Employee was working: ☐ Alone   ☐ With crew or fellow worker   ☐ Other, specify _____

21. POSTURE OF EMPLOYEE AT TIME OF ACCIDENT
☐ Standing   ☐ Kneeling   ☐ Lifting
☐ Sitting   ☐ Walking   ☐ Other, specify
☐ Bending   ☐ Stooping   _____

22. SUPERVISION AT TIME OF ACCIDENT
☐ Directly supervised   ☐ Not supervised
☐ Indirectly supervised   ☐ Supervision not feasible

FIGURE 4-7. Example of an Accident Report Form (side A) for use in recording the necessary general and specific information pertaining to the accident event.

ACCIDENT REPORT FORM

SIDE B

Case or File Number	CAPITAL AUTOMATION INCORPORATED ANYTOWN, TEXAS 12345	Date This Report

SPECIFIC INFORMATION

23. DESCRIBE HOW THE ACCIDENT OCCURRED

24. CAUSAL FACTORS. Events and conditions that contributed to the accident.

25. CORRECTIVE ACTIONS. Those that have been, or will be taken to prevent recurrence.

APPROVALS

PREPARED BY	DATE: / /	APPROVED BY	DATE: / /
NAME:		NAME:	
TITLE:		TITLE:	
SIGNATURE:		SIGNATURE:	

FIGURE 4-8. Example of an Accident Report Form (side B) for use in recording the remainder of the specific information pertaining to the accident event.

etc.) deemed necessary by the investigation team. All of the information used to eventually conclude the investigation will become part of the final investigation report.

The final report itself is usually written in a narrative format, in simple, non-technical language. Figure 4-9 shows a recommended outline for a typical Accident Investigation Report. Such a format offers simple and understandable structure to the report and allows the reader to progress through the sequence of the investigation to its conclusion. It is important that the final report be complete, with logical conclusions and justifiable recommended actions. Because so many different organizational elements will probably have access to the document, it is also important that the organization's legal counsel review its content *prior* to release.

Using the outline presented in Figure 4-9, the accident investigation report should provide at least the following essential information:

1. Introduction: This section is somewhat self-explanatory. It briefly states the *purpose* of the investigation and establishes the scope of the report. For example, in the introductory paragraphs, it is common to affirm that the sole purpose of the investigation is to determine the cause or causes of the accident event and to ensure corrective actions will be implemented to prevent future, similar losses. The *scope* defines the extent of the investigation (i.e., whether it was an all-encompassing effort or a focused evaluation of a specific work area or task). The nature of the accident event and the degree of losses incurred will usually determine the scope of the investigation required.

2. Methodology: For the purpose of documentation, it is recommended that the accident report list and explain the methods and techniques used during the investigation. This will not only serve to document an essential aspect of the investigation, it will also assist the reader in visualizing its extent and thoroughness. The reader should know that the investigation team utilized techniques such as engineering and metallurgical failure analyses, for example. If system safety techniques were employed, it should be stated in this section of the report. Particular analytical equipment such as noise-level dosimeters, air-sampling devices, or light-level meters, as well as any external analytical laboratories used to evaluate the results of any such sampling, should also be documented here. Listing of equipment make and model numbers, plus their calibration dates, will help eliminate the potential for future questions about the validity of sampling results, especially during litigation proceedings. Some investigators prefer to include such information in an appendix to the report. This is also an acceptable approach. The specific location of this information in the report is not as important as ensuring it is included. It is entirely up to the writer. Placing it up-front will allow the reader the advantage of knowing how the investiga-

ACCIDENT INVESTIGATION REPORT

I. **INTRODUCTION**
   A. Purpose of Investigation
   B. Scope of Investigation

II. **METHODOLOGY**
   A. Techniques & Methods Used
   B. Sources of Expertise

III. **SUMMARY OF ACCIDENT**
   A. Brief Description of Accident Event
   B. Statement of Losses & Injuries Incurred

IV. **INVESTIGATION BOARD MEMBERS**
   A. Listing of Board Members
   B. Selection Criteria

V. **DETAILED NARRATIVE OF ACCIDENT EVENT**
   A. Explanation of All Known Facts
   B. Description of Three Phases of Accident
      1. Pre-Contact
      2. Contact
      3. Post-Contact

VI. **FINDINGS & RECOMMENDATIONS**
   A. Findings
      1. Determination of Causal Factors
         a. Primary causes
         b. Secondary causes
         c. Contributory causes
   B. Recommendations
      1. Solutions
         A. Corrective action plan

VII. **CONCLUSIONS**
   A. Summarized Description
   B. Re-state Recommended Actions

VIII. **APPENDICES**
   A. Accident Investigation Forms
   B. Schedule of Investigation Activities
   C. Investigation Models and Analyses
   D. Photographic Evidence
   E. Collection of Witness Statements
   F. Maps, Sketches, Drawings
   G. Acronyms & Abbreviations
   H. Miscellaneous Information

FIGURE 4-9. Suggested outline showing the contents of the written Accident Investigation Report.

tion was pursued before they read the report itself. This may increase understanding of the accident event, an important consideration for the reader attempting to understand the investigation, findings, and recommended corrective actions.

   3. Summary of the Accident: Before the reader progresses into the details of the report, it is helpful if a very brief and simple summary of the accident event is provided. A short paragraph stating the simple facts of the accident—such as the date, time, and location, with a summary of the extent of losses (injuries, property damage)—should suffice. This section draws no conclusions as to cause and does not list any of the recommended corrective actions. It is merely intended to acquaint the reader with the basic facts of the accident event and to set the mode of the investigation that followed.

   4. Investigation Board Members: Before the detailed report begins, it is important to list the names of those persons who served as members of the investigation team. Starting with the chairman, each member should be listed with a brief explanation of his or her background and responsibilities during the investigation. The criteria used for selection of the team should also be documented here. For example, some members may have been selected because of their technical, managerial, or analytical skills in a certain area applicable to the accident situation.

   5. Detailed Narrative of Accident Event: This is obviously the body of the entire report. All the known details of the accident event are explained here in simple and understandable terms. The chronology of events that led up to the accident (the pre-contact phase), those that occurred as the accident transpired (the contact phase), and the events that followed (the post-contact phase) are all documented and explained in this section. A complete listing and description of the losses is provided, including names of injured parties and the degree of their injuries. The names of witnesses are documented here as well. All evidence uncovered during the investigation, whether physical or informational, must be accurately described in this section. Results of any and all analytical evaluations are provided along with the explanations of each by the appropriate investigation team members. While this section does not list findings, nor recommend solutions, it does provide the reader with a clear direction toward accident cause. As each phase of the investigation is explained in detail, the reader will begin to form conclusions based upon the information presented in this section. Therefore, if this section is properly written and presented, then there should be no real surprises when the findings (i.e., causes) are presented in the next section of the report.

   6. Findings and Recommendations: Since this section fulfills the purpose of the investigation report, it is established as a separate section from the previous *(Detailed Narrative)* and the next *(Conclusions).* Here, the investigation team's determination of all causal factors is presented and explained. Once again, if the previous section is properly written, the explanation of the findings should not require much detail. The primary causal factors should be somewhat obvious at this point in the report. In some cases, however, additional elaboration may be required, especially when multiple causes are considered to be primary factors. When secondary and contributory factors are also identified, the writer must ensure the report clearly justifies each causal factor and its relationship to the accident event. Speculation, opinions, and assumptions hinder the clarity of the findings and subsequently decrease the overall validity and usefulness of the investigation. Writers must ensure that the report remains a clear presentation of the facts.

The second half of this section provides recommended solutions or corrective actions that, if implemented, should alleviate the potential for a recurrence of the loss event. Defining solutions is not always as easy as identifying problems. However, the report will not be considered complete unless recommended corrective actions are provided to management. This should include an estimate of total cost of implementation as well as alternate actions should the original suggestions prove infeasible or impractical to implement in actual practice. Providing upper management with several choices is always more successful than simply giving them an ultimatum. After all, if the primary purpose is to prevent loss, there is usually more than one way to achieve that goal. The astute accident investigator should be able to accomplish the objective of accident investigation and loss control without sacrificing the quality of any corrective action plans.

On the subject of recommended actions, it should be noted that management has the ultimate responsibility to accept or decline any recommendations provided in the report. It is a management prerogative to incur the risks which may result from inaction or inappropriate action. The investigation report simply provides recommendations, based upon a detailed analysis of the loss event, that will reduce or eliminate the probability of a recurrence.

7. Conclusions: As a means to an end, this section wraps up the investigation report by providing a summarized description of the accident, including a brief recap of the stated findings and recommendations. At this point in the report, there should be no aspects of the investigation left unclear. However, if this is not the case, the *Conclusions* section should be used to provide final closure and/or clarification of uncertainties.

8. Appendices: Following the outline in Figure 4-9, the information contained in the *Appendices* section of the report should be self-explanatory.

Management Review and Approval

Before the completed report is released and after it has been evaluated by appropriate legal representatives, the final report should be forwarded to upper management for their final review and approval. The primary purpose of this step is for management to *validate* the adequacy of the report, its findings, recommended corrective action plan, and the overall investigation itself. It must be remembered that the function of loss control is primarily a management responsibility. Therefore, any report that discusses a past loss event and provides recommendations to control future losses must be reviewed, and any recommendations approved, by the appropriate level of management *before* implementation of corrective actions. In some instances, upper management may elect to have the company's safety and health professionals examine the document. This step would provide an additional check-and-balance process into the overall investigation. Whatever the final course of action chosen, the choice must be a *management decision*.

DOCUMENTING LESSONS LEARNED

In terms of loss control, perhaps one of the most vital elements of the investigation process is the identification of those lessons which have been learned as a result of having suffered an accident experience. Without exception, accidents are the direct or indirect result of some series of inappropriate actions or inactions. As these are identified, examined, and labeled as causal factors, they should also be documented as lessons learned so that others may benefit from this knowledge. Documenting the lessons learned should include a very brief explanation of the circumstances about the accident, followed by a listing of the actions or inactions which contributed to the resulting loss. Some typical examples of lessons learned may include the following:

1. Improperly trained personnel operating equipment;
2. Operating procedures unclear or otherwise inadequate;
3. Inappropriate tools used for the job or task;
4. Wrong materials purchased due to ordering errors;
5. Manufacturing defect in equipment causing premature failure;
6. Lighting in work area inadequate for proper inspection.

After having defined the primary, secondary, and contributory causes of a particular accident event, it is incumbent upon the investigation team to ensure the appropriate dissemination of the lessons learned throughout the organization and, in some cases, industry as well (i.e., *GIDEP* Safe-Alerts). It should be noted here that the proper documentation of lessons learned is not considered complete unless the success of any implemented corrective actions is also provided. In some cases, the corrective actions implemented after an accident investigation prove to be ineffective at preventing recurrence. This too is a lesson to be learned and must therefore be properly documented as well.

LEGAL ASPECTS OF ACCIDENT INVESTIGATION

Throughout this text, the importance of legal considerations pertaining to accidents, losses, and the resulting investigation of each, have been mentioned time and again. The ramifications of a faulty investigation are serious, not only in terms of accident recurrence, but also with regard to any resulting litigation. During any ensuing legal proceedings, a faulty investigation or an improperly prepared final report can surely be used against the company and its management team.

Investigators, being representatives of management first and accident investigators second, must not only satisfy the objectives of the safety, loss control, and accident prevention program, they must also ensure that liability and the possibility of future litigation are kept to an absolute minimum. This becomes difficult when attempting to isolate accident cause, since most accidents result from a breakdown in the management system. From a purely legal perspective, the entire process of accident investigation can be self-incriminating for the investigating manager and the company as well.

This is why internal legal review becomes an essential element of the investigation process. For example, when the final report contains language indicating that a particular company product or process was defective or dangerous, the courts may view this as an admission of guilt admissible as evidence against the company itself. In an adversarial legal system such as exists in the United States, the "simple truth" is not always so simple. *Truth* in court is based on the ability of each side involved in a litigation to develop its respective position based upon the facts presented during the proceedings, as well as the ability to successfully dispute those presented by the opposition. If the investigation of the accident is determined to be flawed or otherwise incomplete, the actual *truth* of the matter can be successfully argued by opposing legal counsel. The typical investigator is not familiar with such concepts since the purpose of investigating an accident is to determine cause and *not* to assess fault or blame. In contrast, the sole purpose and intent of litigation in such instances *is* to determine fault and assess blame. Obviously, with such diametrically opposed objectives, the investigation report, concerned with establishing one kind of *truth,* can actually be damning in the hands of a plaintiff's attorney who is searching for a different, more profitable version of the *truth.*

The only way the investigator can hope to reduce the level of potential liability is to conduct the investigation with the focused integrity such situations warrant. The fact of the matter is that an accident, no matter how minor, is a litigious situation. If the investigator can remain focused on providing only the most accurate investigation of causal factors, with full consideration of the legal aspects throughout the process, then the risk of making potentially self-destructive errors *may be* substantially reduced. Except where the concept of *attorney-client privilege* has been employed, virtually every document generated during the investigation process (e.g., witness statements, photographs, maps, sketches, drawings, measurements, testing results, and so on) are "discoverable" during legal proceedings. The investigator must therefore ensure that each document is accurate, clear, and complete. Only the minimum amount of information necessary to document the facts should be presented. With the possible exception of witness statements, all necessary precautions should be exercised to exclude opinion, speculation, and assumption from any and all documents related to the investigation. Even the witness statements, if properly solicited, should be void of most opinion and conjecture, with concentration only on the facts. However, it is understandable that keeping a witness from philoso-phizing is not an easy task.

Throughout the investigation, the legality of all actions taken should be considered. During each of the following investigative activities, the actions listed would help to decrease the potential liabilities:

1. Arrival at the Scene: The area must be secured and made safe. The injured must be removed from further danger and their injuries attended to by medical personnel as soon as possible or practical after the accident. Precautions to protect physical evidence from further damage or removal from the scene must be implemented immediately. Witnesses must be identified and documented before people have a chance to leave the area. In extremely active work areas, all possible effort must be made to docu-ment the position of equipment and materials, the procedures in use at the time of the accident, the location of all people involved, and the specific condition of the work environment before they are allowed to change. In short, the company must be able to demonstrate that it immediately took all reasonable steps to minimize the extent of injury and damage resulting from the accident as well as to secure the scene in preparation for the investigation. Such actions fulfill two primary objectives: The investigation will be better served, and any subsequent allegations of inadequate response will be defensible if actions were properly performed and documented.

2. Preservation of Evidence: It is the responsibility of the investigator to ensure that all physical evidence is properly preserved, labeled, or otherwise documented. The location and identity of each piece of evidence collected at the scene, as well as the informational evidence obtained as a result of witness interviews, statements, and the like, must be strictly controlled. The *chain of custody* for each piece of evidence must also be properly documented, from the time it is collected until the conclusion of the investigation and on into any pending litigation. Mislabeled evidence, improperly recorded statements, lost analytical results, and so on, not only reveal a sloppy investigation that hinders the overall process, but also serve as proof to a judge and jury that the subject investigation was flawed. The competency of the investigator may even be questioned. The attorneys for the plaintiff will surely use every shortcoming of the investigation to the advantage of their client. It is therefore incumbent upon the investigator to ensure that no such shortcomings exist that can be turned against the company in a court of law.

Also, in some cases, important evidence from an accident event may not be easily accessible for the investigation team. If all the evidence is within the boundaries of company-owned property, then evidence collection, examination, and preservation is generally not remarkably difficult. However, when the evidence happens to be on the property of another party, there are certain legalities concerned with access which must be considered. Such access becomes especially difficult when the evidence is on the property of an opposing party involved in a lawsuit. Since an investigator will not be able to enter the area without permission of the property owner, the resulting investigation could be somewhat limited. To prevent any possible charges of trespass, the investigator should not proceed further without first obtaining legal permission either from the property owner or the courts. There are certain legal steps which an attorney can take to obtain such permission from the courts to access otherwise inaccessible evidence. Similarly, there are also legal procedures which require the holder of any evidence to properly preserve and protect it while an investigation is underway.

3. Writing the Report: The ability of the courts to access or "discover" the written accident investigation report depends primarily on the status of the investigator at the time of the investigation and at the time of trial (Ferry 1981). For example, a report written by an investigator who is a regular employee of the company and whose job function includes the investigation of accidents, is normally fully discoverable by an opposing attorney. However, when an investigator is specifically requested to perform the investigation in support of a pending litigation, then the resulting report may be protected by privilege of confidentiality between the attorney and the client. Such *work-product* privileges place specific limitations on the ability of opposing counsel to discover certain elements of evidence as the case is being developed. Also, when a client confides in his/her attorney, the courts recognize the confidentiality of such conversations in the interests of ensuring that only the full and truthful facts are being disclosed. Without the establishment of such protection, some individuals may hold back important information if they feel that their words may incriminate certain individuals or themselves. These rules evolved to prevent an attorney of the first party from using the efforts and resulting work product of the attorney of the second party to develop his/her case against the second party. The courts fully expect each side to develop its own case and the rules of discovery, attorney-client privilege, and work-product privilege are therefore quite specific. Hence, when an investigator is brought in to assist an attorney developing a case (i.e., becomes an *agent* of the attorney), the rules of privilege can be imposed to protect against any discovery by the opposition of information obtained by the investigator. Once the attorney-client privilege is invoked, the work developed is protected. However, it should be pointed out that in some cases an investigator may be called as witness to testify in court during the case hearing. Once this happens and an investigator is now a potential witness, all privileges are waived and the opposing attorney has the right to interview and depose the witness. The accident investigation report written by the investigator will also become discoverable at that time. It is therefore important that investigators be conscious of the fact that, even though their words may be protected by the privileges imposed by representing legal counsel, there is always the chance that such protection will be waived as the case proceeds towards its assigned court date.

   4. Public Relations: Often overlooked until it is too late, media and press interest in a given accident can become a serious problem for the investigator and company attorney. In terms of proving liability during case development, innocent statements or opinions made by an investigator or some other company representative, and then made known to the media, can substantially damage the chances for a successful litigation.

   Statements made in the absence of factual evidence are purely hypothetical speculation and should be avoided. Drawing conclusions based solely upon personal opinions and assumptions is an extremely dangerous practice. No one, especially the investigator, should ever speak to or communicate with the public, the press, or the media unless first authorized and designated to do so by a senior company official, along with the company's legal counsel. Most organizations have a designated public relations officer who is skilled, trained, or at least aware of the importance of the proper methods of communication. This is a highly advisable practice since most companies, especially large ones, would find it extremely difficult to control the words of each and every employee who may be contacted by the press for comment. If there is a known, publicized, and established policy within the company regarding contact with the press, then employees will know in advance that only a designated company official is authorized to discuss such matters on behalf of the company. Employees are likely to defer any request for comment to the proper company authority if they are made aware of such a requirement well ahead of time. Employees should also be cautioned to refrain from commenting about an accident to their friends or even to their own families. It is not surprising that a comment about the accident made to a friend at a local bar can become an "official version" of the facts on the evening news or "company representative wishing to remain anonymous" in a local newspaper (Figure 4-10). Once different versions of

FIGURE 4-10.   The proper control and distribution of factual information about an accident is essential to prevent unwanted publicity from the news media.

"the facts" appear in the news, it becomes much more difficult to defend the case during any resulting lawsuits.

When an accident event involves a fatality or receives significant media attention, certain state and federal regulatory agencies are likely to investigate. Certainly in the case of a fatality, the company must notify the proper agencies. However, if the media is reporting inaccurate versions of the facts pertaining to a non-fatal accident event, the regulatory agency may have developed certain impressions and preconceived notions based on this information before they even arrive at the site. While most agencies are familiar with the press and their exceptional ability to distort the truth of the matter, there is still a degree of damage control that will have to be performed once these agencies become involved. It is much better in the long run to control the information being released to the media than to attempt to correct the inaccuracies after the fact.

It is advantageous in some instances to use the press and their ability to rapidly communicate information to the general public. As information regarding the accident becomes known and the facts can be verified, some companies will release information and announcements about the accident in a measured and controlled fashion to the press and local media. Such actions can effectively counter any rumors of hazardous situations, product deficiencies, or other risks to public or employee safety and health. In short, considered efforts to reduce adverse publicity up-front will decrease the amount of damage control required after the investigation is complete or during any subsequent legal proceedings.

5. During Legal Proceedings: When an investigator becomes a witness, there is a likely probability that he or she will be questioned by the opposing attorney prior to the actual court date (i.e., during a discovery deposition). Depending on the perceived value of the testimony obtained during this deposition, the investigator may also be called as a witness to appear in court for questioning in the presence of a judge and jury. It is extremely important that the investigator realize that the opposition, while appearing adversarial, is only attempting to obtain as much information as possible through questioning the facts of the accident. The opposition's job is to prove negligence, fault, tort, or some other degree of liability. Their only method to accomplish this objective is to discover exactly what information exists and then to try to use it to the advantage of the plaintiff. Questioning a witness under oath through deposition, prior to the actual trial, is a form of discovery. During such stressful exercises, investigators must conduct themselves as professionals concerned only with the presentation of fact. They should never loose their patience or otherwise become visibly stressed, nervous, or angry with the attorney's questioning, especially in the presence of a judge and jury, which will only result in lost credibility of the testimony.

Depending on the nature of the case, success or failure may hinge on the confident credibility portrayed by the investigator under examination. Also, when the answer to a particular line of questioning is beyond the knowledge or expertise of the investigator, it is important that the investigator not attempt to provide a response. Simply stating that he or she does not know is much better than providing a guess or an answer that is partially correct or totally incorrect. Sticking to the facts and avoiding hypothesizing will serve both sides well since only the facts will be presented and examined.

While litigation, or even the mere thought of confronting attorneys in general, may be rather intimidating to some people and downright scary to others, the *trained* investigator who is able to maintain a professional attitude, a firm conviction of the facts of the matter, and a visible, confident demeanor should not have any serious problems remaining composed in a litigious environment. It is essential that investigators called as witnesses not appear to be advocates for either side. They should concern themselves solely with the issues of fact as determined during their investigation of the accident. It is when those facts do not reflect favorably on their own company that the manager/investigator becomes entangled in a quandary of professional ethics, loyalty to their company, personal integrity, as well as legal obligation. While such cases (where the company is determined at fault) generally settle out-of-court, such situations still place the investigator in an extremely uncomfortable position. There is no real advise to offer here except to say that those who are unfamiliar with the discovery process will not fare well in the often adversarial legal system within which investigators are sometimes drawn.

SUMMARY

This chapter introduced the concept of *accident response readiness* and discussed the variety of actions that should be taken immediately following the loss event. A primary consideration of the investigation process is to determine the nature of all potential response activities and to adequately account for each during the planning phase. To accomplish the most effective accident response plan, most investigators adjust their readiness planning activities to respond to the *maximum credible event* (i.e., the worst possible outcome of an accident situation) so that anything less catastrophic would certainly be covered. From the initial response action of *notification* through the filing of the accident investigation report, there are numerous opportunities for the investigating manager to ensure that the absolute minimum loss results from the accident. During rescue activities, where care of the injured is the primary concern, the manager can immediately begin loss reduction activities by ensuring that no further injuries occur and having those who have been injured promptly and efficiently removed from further danger. Excessive or additional equipment and property damage can also be avoided during those initial moments of first-response. Steps to *secure, protect* and *preserve* evidence just after the accident, as well as during the impending investigation, must also be taken to prevent loss of critical facts needed to lead the investigator to the proper identification of accident causal factors. This is especially important when the evidence is particularly fragile (i.e., prone to damage, distortion, breakage, and so on). *Fragility of evidence* can be categorized into four major areas of consideration; sometimes referred to as the *four P's of fragility*, the investigator must consider the *people, positions, parts,* and *papers* that were involved in each of the three phases of the accident sequence (i.e., pre-contact, contact, and post-contact).

Also critical to loss control is the investigator's ability to properly communicate with the public. Inadequate public relations, or the lack of it altogether, can result in large-scale distortion of the facts surrounding a loss event being reported to the general public. The potential damage to long-term company business can be irreparable and, albeit indirectly, further the degree of loss incurred as a result of the accident. There is also a certain amount of planning needed to ensure that any notification to the family of an injured employee occur as compassionately and thoughtfully as possible. No family should ever learn that a loved one has been injured (or worse) while listening to the evening news. Companies must not neglect this important aspect of the accident notification process. Assigning a company representative the responsibility of family notification *prior* to any accident occurrence will allow for the accomplishment of the necessary training in human relations and behavior that will be needed to properly and tactfully handle such delicate situations.

Witness interviews are used to gather the *informational evidence* that will serve to validate much of the physical evidence discovered at the scene. Because each person is different and his or her ability to understand, interpret, and articulate what has been seen will vary dramatically, the investigator must learn to approach the witness interview with a basic understanding of human behavior and the learning process. Specifically, the investigator must become familiar with the six *laws of learning* known as readiness, exercise, effect, primacy, intensity, and recency.

Once all evidence has been gathered and evaluated, the investigator compiles the accident investigation report. This report serves to document the adequacy of the investigation as well as provide management with recommended solutions and corrective actions. Since loss control is a management function, the report must be reviewed by the appropriate levels of management prior to its release. Also, because of the extremely adversarial nature of the legal system in the United States, corporate legal counsel should also extensively review the report before it is released.

When an investigator works specifically on behalf of legal counsel to investigate an accident event in support of case development, his or her *work product* can be protected by *attorney-client privilege*. This means that the results of the investigation, including the accident report, are protected from the *discovery process*. However, if the normal job function of the investigator is to investigate accidents, and the information he or she uncovers is documented before the accident becomes the subject of any legal proceedings, then his or her resulting is generally fully discoverable. Also, once an investigator becomes a potential witness who may be called to testify about the investigation, the attorney-client privilege is waived, the investigator can be deposed, and all work products become discoverable. While being drawn into the legal arena is not the most desirable of circumstances, the investigator who has become familiar with the process and can maintain his or her composure, with a firm conviction to the facts, will usually fare quite well during any stressful encounters with opposing counsel. Being prepared for such possibilities is all part of the planning process associated with proper accident response actions.

5

The Human Element

INTRODUCTION

In the previous chapter, *human behavior* and its impact on the overall utility of the witness interview were discussed briefly, to introduce the reader to an extremely important element of the investigation process. This chapter will focus further on human involvement in all phases of an accident event. Specifically, *human error* and *human factor* as accident causes will be explored.

The exact level of human involvement in an accident event is seldom an easy determination, even for the accomplished investigator. However, even though it is often extremely difficult to validate human error as a causal factor without additional evaluation, it is still typically listed as a primary or secondary cause of many accident events. This is especially true in cases where the exact cause of an accident cannot be identified through all available investigation techniques (such is the case with many aircraft accidents, for example). Unfortunately, it sometimes becomes convenient for the investigator to utilize the human error cause when no other can be firmly established. In reality, it can often be proven that a person makes mistakes (errors) as a direct or indirect result of numerous influencing *factors* which may or may not have resulted in an accident (the domino theory).

Upon completion of this chapter, the reader should be capable of approaching the subject of human error in accident investigation with sufficient insight to thoughtfully and competently evaluate the human element in cause determination. When the human factors that directly and indirectly influence the way people behave, react, respond, and perform in a given situation are adequately evaluated, only then can the investigator determine the true nature of human error-caused accidents.

HUMAN ERROR VS. HUMAN FACTOR

Very simply, human error can be described as the *end result* of various factors that acted separately or together to influence the human element during the performance of a particular process. Human error, in and of itself, should therefore never be listed as the sole explanation of cause for an accident or incident. The investigator must look deeper into the cause of such errors. The basic question: What factors, or group of factors, caused a person to make an error? For instance, people may choose an incorrect method or make the wrong decisions based upon poor information provided to them, inaccurate or outdated procedures, lack of proper training, faulty resources (equipment, software), and so on. Resultant "errors" are not actually the true causes of any subsequent accident. The reasons *why* inaccurate information was given, or *why* outdated procedures were being used, or *why* training was not up-to-date, must be thoroughly examined to determine the real cause of the human error.

The use of the term *error* without any designated cause is a subtle assessment of blame or fault that, of course, is neither the objective nor the business of an accident investigation. To understand the term further, *Webster's New World Dictionary* (1988) defines *error* as "the state of believing what is untrue, incorrect or wrong; a wrong belief; incorrect opinion; something incorrectly done through carelessness or a mistake; a sin." *Webster's* also recommends seeing the word *blunder*. Blunder is a more inclusive term, meaning "stupid mistake, and, to err stupidly; to move clumsily or carelessly; flounder; stumble; to do clumsily or poorly; a foolish, stupid mistake." The *Dictionary* further indicates that *error*, in general terms, is any departure from right or truth and that we may commit errors in judgment, or in computation, and that mistakes and blunders are kinds of errors. A mistake is an error of carelessness, inattention, or omission. A blunder is an error caused by stupidity, ignorance, conceit, or failure to understand.

The intent of this vocabulary lesson is to demonstrate that an error is actually caused by a wide variety of definable factors. However, these factors are only definable when *all* possible aspects of the accident cause are considered during the investigation. Decades of accident statistics reveal that little success has been made in dealing with the prevention of human error-caused accidents. Historically, human error has been named as the primary cause in a majority of accidents throughout both the public and private sectors. In fact, these statistics clearly proclaim that the finding of "human error" as the cause of accidents will seldom prevent a recurrence of an accident by the same cause. In other words, a finding of human error today will most likely repeat as a cause factor tomorrow. Evidence indicates that listing human error as cause, without exploring the factors that led a human to err, will have little or no effect on the prevention of future similar events—in direct contradiction of the objective of accident investigation and loss control.

Listing human error as the primary cause of an accident has not been very successful in preventing additional accidents. This may indicate that human error is not really a cause but a *symptom* of a much deeper problem. Technological causes of accidents are usually determinations based upon scientific proof as developed through a detailed examination and evaluation of all physical evidence. However, in most cases, human error causes are typically based upon a combination of conjecture, speculation, assumption, and unproved hypothesis. In fact, human error is usually determined by exception; that is, if the investigation can prove nothing else, and circumstances fit, a convenient finding of human error is subsequently made. This unwritten law of exception seems to imply that, if all possible causal factors have been ruled out except for the human element, then the cause must be human error. This suggests that all causes of accidents must be predicated upon proof, with the exception of human error. This, of course, is poor logic.

Management and Responsibility

A basic premise of effective business operations may be summarized as follows: Every person assigned a specific or general job function should be *responsible* for ensuring that each assignment is completed properly, safely, efficiently, and without incident. However, this does not in any way relieve management of its responsibility for overall safe operations. Long-time managerial authorities Koontz and O'Donnell explained the rules of organizational responsibility three decades ago (Koontz and O'Donnell 1964):

> Responsibility cannot be delegated . . . no manager can shift such responsibility to subordinates. The president cannot avoid total responsibility for the conduct of the enterprise . . . management cannot claim to have delegated the responsibility to a subordinate, who may, indeed, have caused the trouble.

Therefore, by application of this long-standing rule of management, there appears no real justification for the determination of human error as the root cause of an accident. As stated in Chapter 1 of this text, a breakdown or failure in the management system is more typically the primary cause of most accidents today. For instance, most accidents usually involve a repeat cause factor (i.e., one that has been a causal factor in the past). When an accident occurs, it is supposed to be investigated for the purpose of preventing recurrence. The information obtained through investigation should be utilized to prevent future, similar loss events. This is the primary nature of a loss control program. If the cause factor repeats itself, it is an indication that the loss control program has failed. But when the cause factor repeats itself again and again as is the case with most human error causes, the actual cause has automatically become a factor of supervisory/management failure. Management failure should then be identified as the *root cause*, and human error becomes only a symptom of that cause.

PHYSICAL AND PSYCHOLOGICAL ASPECTS

Every day, people are required to execute a variety of simple and complex tasks throughout the performance of their work duties, as well as during the conduct of their private lives. Individuals' talents, abilities, and contributions become the basis of all that is accomplished in every aspect of daily living in our world. So much, in fact, depends upon the successful performance of the human element to ensure the favorable progress of social and economic development. When human performance is somehow less than anticipated, expected, or desired in a given endeavor, the rate of progress is equally hindered. Therefore, it is incumbent upon all those who investigate the causes of human failure to ensure that an adequate evaluation of the factors that affect human performance has occurred.

These *factors* can be divided into two basic categories: physical and psychological. Table 5-1 lists only a few examples of physical and psychological factors that can adversely affect human performance. The manager/investigator must be cognizant of the potential effects such factors can have on an employee's job performance. While a person may indeed make an error that resulted in an accident, the investigation should reveal the actual factors that influenced the employee's behavior and may have resulted in the error. Once these human factors have been accurately identified, it becomes much easier to recommend solutions and corrective actions. In most cases, an investigation that closes with a simple finding of "human error," with no further explanation or discussion, usually reflects an incomplete and therefore unsuccessful accident investigation.

Analyzing the Problem

To better understand how human performance can be affected by adverse physical and psychological factors, such as those listed in Table 5-1, the

TABLE 5-1   Some common examples of physical and psychological factors that can adversely affect human performance.

FACTORS AFFECTING HUMAN PERFORMANCE	
PHYSICAL	PSYCHOLOGICAL
HUNGER	STRESS
THIRST	EMOTIONAL DISTRESS
FATIGUE	PREOCCUPATION
WEAKNESS	LACK OF CONCENTRATION
STRENGTH	LEARNING DISORDERS
INJURY	ATTENTION DEFICIT
HEIGHT & WEIGHT	IGNORANCE
PHYSICAL ILLNESS	FEAR
VISION INADEQUACY	MENTAL ILLNESS
HEARING INADEQUACY	ANXIETY

human element can be evaluated for potential failure, just as a given system or piece of equipment can be similarly examined. In a purely analytical sense, examining the characteristics of the human component in work systems reveals several differences between the human and hardware elements in terms of those factors that affect the expected performance of each. Some primary differences reside in the following areas of consideration (SSDC-32 1975).

The Basic Response Model
When evaluating the capability of the human and the hardware (machines, equipment, etc.) to respond to a given set of commands within an established set of parameters, the basic response of each provides valuable information in terms of accident investigation and analysis. For instance, the hardware element responds largely as a two-step model; that is, a simple stimulus-response reaction. If a relay is wired into a system and voltage is applied to the relay as the stimulus, the relay will either open or close its

contacts. If it has an open coil, it will not respond. But the investigation reveals application of the stimulus either *does* or *does not* cause a response, there are no other possibilities.

In some instances, there are similar responses in the human. If a person's knee is tapped in the proper way, and the neurological circuits are working, the reflex loop will cause the foot to spring forward. But in the work environment, this simple stimulus-response action is not ordinarily used with humans in the system. Generally speaking, if a stimulus is applied to a person, there may be a wide variety of responses other than simply *doing* or *not doing* a specific task.

Unlike the hardware element, a person usually reacts in a multiple-step fashion. First, a person must *receive* the input as it was intended to be received. Any error made at this point could result in a wrong response and a subsequent accident or near miss. Second, he or she must *recognize* or properly interpret the requirements of the input, which may have been received along with numerous other simultaneous inputs. Once again, misinterpretation of intended requirements can lead to undesired results. Third, the person enters a *mediation process* and either responds or does not respond. Similar to the median that separates two sides of a highway, the mediation step lies between recognizing the stimulus and actually responding to it. Unlike a hardware element, a person has the ability to reason, decide, and choose between action and inaction. Hence, the mediation step occurs when the person *thinks* over a decision based upon such *human factors* as his or her prior knowledge and experience, personal goals, emotions, training, and any other aspect of that person's individual behavioral processes. The mediation step has two primary effects on the work process: it introduces delays and it has a potential for producing a wide variety of responses. As an example, if a human operator is given verbal instructions or is exposed to certain indications or displays on a control panel, there are numerous possible responses, ranging from doing exactly what was expected to actually proceeding in a diametrically opposite direction. The forth and final step, *response,* is that in which the human error-caused accident will or will not occur as a result of all those factors that influenced the person during the process. Attempting to correct the outcome at this stage is usually futile since so many inaccuracies probably occurred during the previous three steps. Throughout each step of this basic human response model, falling dominos may eventually lead to an accident.

Quality Assurance

There is a great deal of difference between the human and the machine in what might be referred to as the quality assurance of a system. In general, the electrical and mechanical components of a new system are typically new components. Older systems have components that are regularly serviced, replaced, or maintained to assure the proper function of the system. In fact, regular preventive maintenance and inspection are primary elements of any industrial quality-assurance program. Rigid specifications are typically established, and there is usually a wide variety of testing to assure that the system meets those specifications.

In contrast, a human enters into a work system composed entirely of "used" components. In other words, with the introduction of the human appearance, the system is forced to accept a component (or multiple components) that has been used, misused, and even abused (depending on the individual) for many years before entering the human-hardware system. Furthermore, specifications established for human beings are crude and subjective compared to the rigid requirements placed on hardware and equipment. Hence, selecting humans and putting them into a system is much like selecting hardware from a used hardware vendor. The quality assurance on humans is much less rigorous, defined, specific, and tested than the hardware elements of a system. In short, every person who enters the system is a used and abused component. While it is true that a determining factor in personnel selection is an individual's experience and knowledge, often there is relatively little awareness of what went into that person in terms of use and misuse, highlighting the effect that unknown human factors and human behavior problems may have on the quality of system performance.

Stress Resistance
When evaluating the hardware in a system, the investigator will discover that there are many elements, such as solid-state circuitry, that embody certain stress-inducing (temperature, humidity) conditions that must be controlled within close tolerances. Allowances for stress factors such as temperature, pressure, and vibration can be designed into the total system-operating specifications. Additionally, when multiple components are assembled, they are generally designed to operate together within a specific operating environment.

Unfortunately, humans have a relatively narrow range of conditions under which effective performance can be expected. Drastic fluctuations in temperatures, pressures, noise levels, illumination, vibrations, or other physical stress factors will dramatically affect a person's performance. In addition, there is much more variability in people, both from person to person and within the same person from time to time, in terms of their individual tolerance to such changes. There are also physical, emotional, and mental stress factors within humans that are not experienced by hardware.

Failure Modes

The end result of the many variations in human performance is a wide variety of possible failure modes for the person that are not experienced in the hardware system. For example, when people are fatigued, their error rate may increase substantially compared to their performance when well rested. Other physical factors, listed in Table 5-1, such as hunger, thirst, physical illness, poor vision or hearing, or weakness, all will potentially increase the rate of failure during the performance of a given task. Likewise, psychological factors (anxiety and stress, emotional disturbances, fear or ignorance, preoccupation, and mental illness) will have similarly adverse effects on the desired outcome. When the investigator considers the potential for disaster when one or more of these physical or psychological factors exist at the same time, he or she does not find it difficult to understand how human factors contribute to so many accidents in our world today.

THE SUPERVISOR'S ROLE

In general, supervisors are responsible for the activities that occur in their associated areas. To accomplish the objectives established for them by upper management, as well as to effectively demonstrate their own managerial skills, each supervisor must efficiently utilize a variety of resources provided by the company. These include equipment, materials, supplies, working area, operating specifications and instructions, and, of course, *people*. Of all these resources, people can be the most unpredictable and, therefore, the most difficult to manage.

To ensure the prevention of accidents in their areas, supervisors must maintain a constant awareness of the potentially dangerous situations that can arise during the normal course of business in each working day. Equipment problems must be identified and corrected before their use becomes hazardous. Material specifications must be checked regularly to determine if any changes in chemical composition or the like will threaten employee safety and health. Procedures must be well-written and regularly reviewed to guarantee the safest approach to the project or task. Many other insidious hazards can be present at any given time in the work environment, which, if allowed to go unchecked, could eventually lead to an accident situation. Environmental conditions such as noise, temperature, illumination, vibration, and even the air employees breathe, should all be within safe and tolerable ranges. In addition to meeting production schedules, ensuring first-time quality, maintaining operations within or below allocated budgets, and performing a wide variety of other managerial and administrative duties, the supervisor must also take the necessary measures to ensure an accident-free workplace.

With regard to the human element, there is much that the supervisor can do in the accident prevention arena as well. For example, the interview and selection process provides the first opportunity for the supervisor to reduce the potential for future accidents. By properly selecting, or "fitting," the employee to the job, the likelihood of human errors resulting from lack of knowledge or experience can be nearly eliminated. Once on the job, supervisors must see to it that their employees maintain the required level or proficiency that the job demands. When requirements change or new equipment is introduced, employees must be retrained. Statistically speaking, the lack of training (or the provision of improper training) is one of the most frequently cited factors related to human error-caused accidents.

Since optimum human performance plays such a key role in the success of any business endeavor, the factors that diminish performance should be the focus of serious attention for every supervisor. Each supervisor must be a keen observer, capable of detecting problems or potential conflicts within the assigned work force. For instance, sometimes there are personality conflicts between employees that can adversely affect everyone in the work area. Quick identification and resolution of such situations by the supervisor must occur if the accident risk potential is to be reduced. These types of problems are often not difficult to detect, if the supervisor is trained in such matters. But, when the accident potential lies within the mind of one employee, the supervisor may have little or no immediate indication that any problem exists. Employees who are under stress and anxiety resulting from personal conflicts such as financial or marital problems, difficulties with their children, an illness or death of a loved one, drug or alcohol abuse, emotional instability, mental illness, and so on, may seem to be quite normal based on outward appearances. However, when work output begins to decline, or the quality of production is not what it should be for a given employee, it is time for the supervisor to start asking the right questions. Hopefully the company has an established *Employee Assistance Program (EAP)* to help employees (and supervisors) cope with such unfortunate but all-to-common problems. Whatever the case, if the supervisor does not become aware that something is not quite right with an employee, then the chances of an eventual accident will only increase as the conflict within that employee continues to grow.

In terms of accident causal factors, the investigator must always look beyond the simple conclusion of human error to determine what underlying factors (physical, psychological, environmental, etc.) influenced that employee to the point where an error in judgment was made. Once these factors are identified, the investigator can then ascertain whether or not management could have intervened during the pre-contact phase to prevent the accident. If this is the case, then the true cause of such an accident would be a failure in the management system, with the resultant human error only a symptom of that failure.

SUMMARY

Generally speaking, nearly every accident event can somehow be linked to a failure in desired or expected *human performance*. Even when an accident is the result of equipment failure or metal fatigue, it can still be related to human error since it was undoubtedly a human who designed the equipment and selected the metal components. If the investigation reveals that the equipment design was inadequate or the parts used incapable of performing within design specifications, then the resultant findings (i.e., accident cause) might list equipment failure as a primary cause and human error as a secondary or contributory cause. While such findings might be acceptable to many organizations, the investigation would not be complete unless the reasons or *factors* that led to the human error are also fully explored. The investigator should attempt to determine the *reasons why* a person selected the wrong equipment or materials. Perhaps the information provided to him or her was incorrect. Or maybe the selection of equipment and materials was beyond the level of an individual's training and expertise. Whatever the reason, the investigator must realize that a simple finding of human error, without further examination of those factors that directly or indirectly influence human performance, is an *incomplete assessment* of accident cause.

Factors that affect human performance can basically be divided into two categories: *physical* and *psychological*. The intent of this chapter was to introduce the reader to the complex issue of human error-caused accidents and how these many factors must be investigated to determine their impact on human performance. More importantly, the *management system* must also consider these factors during the course of daily operations to ensure an accident-free work environment. In addition to managing a budget, working to production schedules, reacting to changes in customer specifications, ensuring the total quality of products and services, allocating required resources, and maintaining a safe and healthy working environment, the manager/supervisor must be cognizant of any changes in employee performance that may or may not indicate the presence of one or more factors that, if allowed to continue unchecked, may lead to a loss event. In many instances, a failure in human performance can be directly or indirectly linked to a failure in the management system. In fact, a simple finding of human error without further discussion will seldom prevent the occurrence of future similar events caused by the same "errors." An accident cause that appears repeatedly is a direct indication of a failure in management's ability to *identify and control* those factors causing accidents within their enterprise. Simply stated, if damaged equipment is determined to be the cause of an accident and management does not replace or repair it prior to placing it back into service, then management has failed to prevent the likely recurrence of the accident event. Likewise, if human error has been determined to be a cause of an accident and management does not adequately address and resolve those factors that led to the undesirable human performance, then management has once again failed to prevent a recurrence. In most cases, human error is not really a cause at all; it is a *symptom* of a much deeper and insidious problem that has its roots in management's failure to recognize and identify problem factors *before* they lead to the loss event.

The investigator, who is also typically a manager or supervisor, must therefore be aware of the potential *shortcomings* associated with a finding of human error as accident cause. Without proper consideration of the human element in the investigation process, as discussed in this chapter, the successful conclusion of the entire accident investigation and loss control effort will be questionable. Remember, the primary objectives of accident investigation are to *prevent recurrence* and to control loss. Managers who do not adequately examine the human element, during the investigation of one accident event, may run the risk of becoming a cause factor themselves during some future investigation of a repeat accident.

II

Tools and Techniques for Investigation

INTRODUCTION TO PART II

Part I of the *Basic Guide to Accident Investigation and Loss Control* provided general discussion on the principles of investigation, the responsibilities of the investigator, the nature of accidents and their causes, the principle elements of investigation planning, the many accident response actions necessary to ensure maximum reduction of loss, and the importance of considering the human element during the entire investigation process. At this point, the reader should have a fundamental understanding of the investigator's role in the loss control program.

Part II will expose the reader to some of the many tools and techniques used by today's accident investigators as they attempt to comply with the objective of the investigation process. As established in Part I, the investigation of accident events, usually performed by management, is concerned with the identification of all possible causal factors and the provision of recommended solutions and corrective actions to prevent the potential for recurrence. In Part II, investigation kits, photography, maps, sketches, drawings, and even system safety techniques, which can be used to facilitate the investigation of loss events, will be discussed. In addition to the tools that are presented in the following chapters, Appendix C of this text provides an example of an accident investigation checklist which can be used as yet another tool to *initiate* the investigation process.

At this point, it is important to note that the availability of these tools and techniques does not automatically equal an effective investigation or loss control program. The proper or improper use of each can be a key element in the success or failure of any accident investigation.

6

The Accident Investigation Kit

INTRODUCTION

As discussed throughout Chapter 3 (Investigation Planning and Preparation), the success or failure of an accident investigation may be contingent upon adequate and thorough preparation. As part of this most important planning process, one of the first considerations that should be addressed is the preparation of the *accident investigation kit.* Having the proper items already assembled prior to the occurrence of any accident will certainly facilitate the initiation as well as the progress of impending investigation activities. For example, there are several items that can be considered generic to any accident investigation. These may include measures (tapes, rulers, etc.), notepads, writing materials, graph paper, and report forms. While it may seem somewhat unnecessary to worry about the availability of such things before any accident has actually occurred, the experienced investigator knows that having even the simplest of tasks completed well ahead of time will greatly facilitate the beginning of any investigation. The investigation kit is just one of many *tools* available to the manager who is called upon to investigate an accident. Depending upon the nature of a particular loss event, the adequacy of the kit could affect the positive outcome of the investigation.

This chapter will briefly discuss the factors that affect the selection of the items to be placed into an investigation kit, the control of the kits once they are assembled, and the use of the items to facilitate the investigation.

INVESTIGATION KIT ITEM SELECTION

The items selected for an investigation kit can be divided into two basic categories: those that will generally be needed for almost every investigation and those specific to the types of loss situations that may be unique to a particular task or work operation. The former may be relatively easy to compile while the latter require some knowledge of the work operations that may be subject to investigation at some future point in time. Having an area manager or supervisor assist in the development of a kit will not only ensure that the proper and essential items are obtained, it will also enlist the early participation of a manager in the investigation process who, in all likelihood, will be the one performing the investigation of any accidents occurring in his or her area of responsibility. Obtaining this early *"buy-in"* of the investigating manager provides on-the-job-training and also develops a sense of accountability for the success of any impending investigations.

Table 6-1 lists the items, by category, that should normally be considered when preparing an accident investigation kit. However, it should be noted that Table 6-1 is by no means intended to be an all-inclusive listing. When fully compiled, the kit must be packaged in a manner that is readily portable. Each item in the kit should be accessible without having to unpack or re-pack other items. To control the size and portability of each kit, the special equipment (cameras and other valuable equipment) that might be needed for unique accident situations should be kept in a separate, smaller kit in which they can also be better protected and controlled. The kits should be stored in a location where they can be easily accessed in the event of an accident and, at the same time, where they will be protected from their tempting use as a routine source of supplies and tools.

Kits may be centrally located in areas such as the safety department (provided that access is still possible during off-hours, holidays, and weekends), or, more practically, in direct control of supervision in those areas where the kits will most likely be needed. A periodic inventory should be performed to check that the items in the kit have not been removed without authorization. A listing of the items contained in each kit could be mounted on its exterior to facilitate inventory of the kit's contents. Items that are subject to deterioration, such as ink pens, adhesives, and some tapes, should be replaced regularly to ensure the kit items remain usable at all times.

While many of the suggested items appearing in Table 6-1 are somewhat self-explanatory, the following information is provided to clarify the potential use of some of these items during an accident investigation.

GENERAL

ADMINISTRATIVE
- Investigator's workbook
- Clipboard
- Notepads, lined, 8.5 x 11
- Plastic zip-lock baggies
- Graph paper, 1/4 " sqrs
- Accident report forms
- Witness interview forms
- Large envelopes, 9 x 12

- Ballpoint pens
- Soft lead pencils
- Grease pencils, black
- Spray paint, orange
- Aluminum foil, one roll
- Paper towels, one roll
- Adhesive labels, assorted
- Permanent markers

- Eraser, lead
- Eraser, ink
- Adhesive tape
- Masking tape, 2 inch
- Cardboard tags, string ties
- Plastic zip-lock ties
- Pocket calculator
- Spare calculator batteries

TOOLS
- Screws, assorted types
- Scissors
- Pliers
- Adjustable wrench set
- Open box wrench set
- Wire cutting pliers
- Saw, wood/metal cutting
- Compass
- Small pick-ax & shovel

- Magnifying glass
- Inspection mirrors
- Toothbrush, soft
- Nails, assorted sizes
- Protractor
- Vice grip
- Screwdrivers, assorted
- Knife
- Hammer

- Awl or icepick
- Duct tape
- Rope, nylon, 50-ft.
- Twine string, 300 feet
- Flashlight, expl. proof
- Spare flashlight batteries
- Tape measure, 100 ft.
- Scale, 12 in. ruler
- Small plastic bucket

MEDICAL
- Water
- Triangular bandages
- Adhesive tape, 1 inch

- Peroxide
- Plastic bandage strips
- Alcohol

- Antiseptic cream
- Waterless hand cleaner
- Eyedrops & ointment

PERSONAL PROTECTION
- Hard hat
- Safety glasses, plano
- Chemical goggles
- Gloves, leather
- Rubber boots

- Gloves, canvas
- Gloves, chemical
- Foam ear plugs
- Vest, orange
- Rubber apron

- Face shield
- Respirators, various sizes
- Organic vapor cartridges
- Acid gas cartridges
- Disposable suit

SPECIFIC

EQUIPMENT
- Camera, 35mm, slide
- Camera, 35mm print
- Polaroid® camera, film
- Micro-cassette recorder
- Extra cassette tapes
- Sound level meter
- Gas vapor analyzer
- Video recorder
- Electrical receptacle tester
- Calipers
- Temperature meter
- Air flow velocity meter
- Light-level meter
- Torque wrench
- Vibration meter
- Synthetic smoke (air flow tester)
- pH meter or paper
- Orange road cones

DEVICES
- Engineer's scale
- Metric conversions
- Wire rope size calculator
- Fluid sample containers
- Pin marking flags, 18 in.
- Road flares
- Syringe, various sizes

DOCUMENTATION
- Equipment manuals, specific
- Emergency procedures, specific

TABLE 6-1   Recommended components of the accident investigation kit.

Administrative Supplies

The *investigator's workbook* should be used to document and categorize the collection of evidence. It should also contain any completed checklists, accident investigation forms, witness statements, graph paper for mapping or sketching exercises, photo logs, sequence diagrams or charts, and any other paper documentation applicable to the investigation of the accident.

A *clipboard* is recommended to facilitate the performance of any field calculations or sketches as well as provide a method to organize notes and written observations made during the investigation process.

Plastic zip-lock bags provide a variety of uses. If large enough, one bag can serve as a rain cover for the investigator's clipboard to permit record-making during inclement weather. Zip-locks can also provide protection and separation of documentation collected during the course of the investigation. Smaller bags can be used to collect and contain parts of evidence that must be removed from the scene for further analysis.

Assorted *envelopes* (varying sizes) can also be used to organize evidence-collection. They provide an increased level of security against intermixing of documents and the loss of pages.

Aluminum foil provides a convenient and inexpensive material to wrap and protect certain parts from further exposure to the elements, thereby decreasing the potential for corrosion and contamination.

Paper towels can be used to pad fragile parts before wrapping to further protect fracture zones or mechanisms during transport for analysis. Absorbent paper towels are also useful during preliminary cleaning and drying of parts and equipment.

Adhesive labels (assorted sizes) can be used to properly identify parts wrapped and secured at the accident scene. Labels provide for quick identification without having to disturb the packing materials to determine the contents of a given package or envelop. Labels can also be used to temporarily identify materials or items being photographed or to highlight special points of interest.

Grease pencils are useful for marking floors, surfaces, equipment, or materials to identify positions or reference points. They can also be used to mark equipment before disassembly to facilitate the re-assembly of the parts as originally positioned. A grease pencil will mark wet surfaces and rough textures much better than chalk.

Orange spray paint allows a highly visible marking on surfaces such as dirt, snow, roads, and other surfaces not suitable for marking with a grease pencil. Caution in its use should be exercised since such markings are much more permanent than those made by some other means. The spray paint can also be useful in placing certain identifying marks on large, damaged structures to be photographed or to outline the position of parts prior to their removal from the accident scene.

Cardboard tags with string ties can be used to label small parts and items when such labeling is more suitable than adhesive labels. It is suggested that a grease pencil be used to label parts in the field so that the information will not be erased or smeared by contact with water.

Tools

With few exceptions, nearly every accident investigation will require the use of a *flashlight* during some aspect of the investigation process. Each kit should include a reliable, explosion-proof, lightweight (preferably a floating type), waterproof, high-intensity flashlight. Spare batteries and at least one replacement bulb should also be kept in the kit. The exterior flashlight casing should be plastic (non-sparking and non-conductive) instead of metal.

Another indispensable tool at an accident investigation site is a *tape measure*. A well-made, durable steel tape measure, with at least one hundred feet (30.5 meters) of tape, can be used to record a wide variety of information critical to the investigation. Modern tapes usually include both standard U. S. customary units (inches, feet, yards) and their metric equivalents (millimeters, centimeters, meters).

Having a straight-edge ruler or *protractor* (12-inch ruled) readily available facilitates the mapping and sketching of accident scenes as well as the measuring of small pieces of evidence. It can also be used as a size reference when taking certain close-up photographs at the accident site.

Assorted *inspection mirrors* will allow for examination of areas not readily visible due to their location or size. Large mechanic's-type mirrors and small dentist's-type mirrors are excellent tools for the investigation of accidents involving complex machinery or equipment.

A *toothbrush* with natural, soft bristles is ideal for cleaning metal fracture surfaces and small parts. Nylon-bristle toothbrushes should not be used because they may scratch or score softer metals such as brass and other soft materials.

Having an assortment of large and small *nails* in the kit (and, of course, a hammer) will allow the investigator to mark positions in dirt or on roads, to hold tape measures, and to act as anchor points for string used to outline debris fields. Small parts can be held in position with small nails during reconstruction activities.

Rope and *string* can serve a variety of useful purpose. For example, the rope can be used to establish a controlled area around the accident site or to section off certain areas where evidence will be collected and examined.

String can be used to tie materials together for reconstruction or analysis at a later time, or to wrap parts for preservation of evidence.

Special Equipment and Devices

Cameras are indispensable tools for the modern accident investigator. It is suggested that different types of cameras be readily available for use, with an adequate supply of film for each camera also kept on hand. Since Polaroid™ cameras provide an immediate photograph of the subject, they are very useful when setting up a photo layout of the accident area, a particular piece of evidence, or a general view of a specific item of interest. A good-quality 35mm camera loaded with slide film is useful for document- ing evidence that may be used during any subsequent presentations to interested parties as well as during any legal proceedings. A similar camera with print film will provide photographic evidence to be included in the documentation and final investigation report. Such photographs also assist the investigator during the analysis phase of the investigation. While it is true that slides can be turned into prints and vice versa, such conversions usually take time. Having at the accident site two separate cameras with both types of film is recommended, since the use of more than one camera increases the likelihood of filming valuable evidence that might otherwise be missed. A video camera offers even greater diversity in the documenta- tion process. In any case, it is highly recommended that the film for each type of camera be kept in a cool, dry place (usually not in the investigation kit) until it is needed. Depending upon the ability to assure kit security and integrity, it may also be advisable to store any cameras in a location other than inside the investigation kits. Additional information about the use of cameras and photography during an investigation is presented in more detail in Chapter 7 (Photography in Accident Investigation).

The use of a *micro-cassette tape recorder* can be extremely advantageous for accurately recording on-scene observations, especially during situations where note-taking might be inappropriate or impractical. Dictation of viewpoints and opinions onto a micro-cassette tape, for transcription at a later date, not only saves time and effort, it also allows the investigator to express thoughts as they occur. Of course, it is just as important to have a good supply of replacement cassette tapes and batteries as it is to have the recorder itself.

Depending upon the nature of the accident being investigated, a *sound level meter* (dosimeter) will be needed to measure noise levels in the accident area. Proper measurement and analysis of sound levels may be particularly critical in determining the primary, secondary, and contributory causes of some accident events.

Certain accident investigations may require the use of a *gas and vapor analyzer*. Insidious gases such as carbon monoxide and nitrogen, as well as obvious gases such as chlorine and ammonia, are common in many of today's industrial applications. The rapid and accurate analysis of airborne concentrations of the types of gases that may be present in a facility could provide the evidence needed to identify the root cause. The investigator must be trained in the proper use of these devices and in the methods used to interpret and evaluate the data collected as a result of any such analyses.

Accidents involving electrical failures, overloads, or other related causes rank among the most common of incidents throughout industry. An inexpensive *electrical receptacle tester* allows for quick and accurate testing of the electrical continuity and grounding adequacy of receptacles and distribution cords. No investigation kit should be without this extremely useful device.

Calipers are used to make precise measurement of parts. It is essential that calipers capable of measuring both inside and outside dimensions be included in the kit.

In most cases, machinery and equipment must use certain types of fluids to operate properly. The appropriate type, grade, weight, or amount of fluids such as fuels, lubricants, hydraulic oil, coolants, or even water is necessary for many machines to operate as designed. *Fluid sample containers* of the type, material, and size needed to collect such fluids for sampling and analysis should therefore be included in the investigation kit.

Synthetic smoke is very useful (and quite harmless) for obtaining a quick visible reference of airflow direction and speed in a given area. It may also indicate the presence of a draft or an otherwise unknown leak in a system designed to be air-tight.

Accident investigations that involve an evaluation of specific equipment or machinery should also include a review of any specific *documentation* associated with the operation of that equipment. Investigators often find that inadequate operating procedures contributed to a resultant loss event. Also, having operating procedures readily available for the investigator at the scene will greatly facilitate the investigation of the equipment as well as provide the correct procedures for shutting down that equipment, as the case may be. Of course it may not always be practical to include a complete operating or emergency procedures document for every piece of equipment in every investigation kit; for some facilities, such a task would require a complete library in each investigation kit. What is suggested is to ensure that any such specific documentation be kept centrally located and readily available to the investigator. When accidents involving specific equipment do occur, the necessary documentation can then be obtained quickly and included in the kit at the beginning of the investigation. Once again, simple preparation is the key to a successful investigation.

ADDITIONAL INFORMATION

Investigators should always be prepared for the conditions in which they may have to work at the accident scene. Wearing clothing that provides suitable protection against any hazardous elements they might encounter, should be self-evident, but it is often overlooked until arrival at the scene. Proper preparation to ensure protection against insidious hazards such as hypothermia, exposure, frostbite, or heat exhaustion, is an essential consideration for the astute investigator.

Also, depending upon the nature and complexity of the investigation, inadequate clothing may prompt an investigator to hurry through many critical investigation actions. When the accident site is somewhat remote or isolated, requiring an extended visit by the investigator, then he or she should be prepared with clothing suitable to the conditions likely to be encountered. Extra garments should also be brought along so that clothing that becomes wet, soiled, damaged, or otherwise unwearable, can be changed quickly, without too much disruption to the investigation itself. While this information may seem somewhat basic, it is important to note that the obvious is often overlooked during initial response actions following an accident event.

SUMMARY

Of all the tools available to the modern accident investigator, none is perhaps more simple to prepare and more valuable to the user than the *accident investigation kit*. This chapter has discussed briefly the development of the kit and presented a list of recommended items that have proven useful during the investigation of accident events.

Basically, the items that should be considered for inclusion in an investigation kit can be divided into the categories of *general* and *specific*. Those listed as *general* usually have some practical utility during most investigations, regardless of severity or complexity. General items and materials can be further divided into four subgroups to simplify the selection process. *Administrative supplies* may include such mundane but useful articles as plastic zip-lock bags, adhesive labels, paper towels, pens, pencils, graph paper, envelopes, and so on. *Tools,* such as screwdrivers, a hammer, some rope, a flashlight, and a magnifying glass, to name only a few, are helpful during the investigation of nearly every type of accident event. *Medical supplies* are also an essential consideration for every investigation kit. Basic items such as bandages, adhesive tape, peroxide, alcohol, and the like, could keep minor cuts and abrasions from adding further to the overall losses incurred as a result of the accident event being investigated. *Personal protective equipment (PPE)*, like hard hats, safety glasses, chemical goggles and a face shield, foam earplugs, as well as assorted gloves (leather, canvas, rubber), and standard body protection (rubber apron and disposable suits), should all be readily available to the investigator in the kit.

Specific items in the investigation kit are further divided into three subgroups. *Equipment* that may be required to properly examine certain aspects of a particular accident event includes special cameras and voice-recording equipment, testing instrumentation noise, light, temperature, and vibration level meters), pH indicators, and an electrical receptacle tester. *Devices* include specialized items such as road flares, an engineer's scale, a wire rope size calculator, or any other device so specific to a particular type of accident situation that inclusion in the kit becomes necessary only when their potential use is obvious or anticipated. It is best simply to have these items readily available rather than kept in a kit for an indefinite period of time. *Specific documentation, such as operating manuals,* that may be necessary to properly evaluate certain equipment or to better understand the emergency procedures used during the post-contact phase of the accident, should be readily available and, if necessary, placed into the kit at the beginning of a particular investigation.

In all cases, the accident investigation kit is a vital aspect of the preparation process. During the often chaotic moments immediately following the occurrence of an accident, it may not be possible to take the time necessary to assemble the appropriate items needed for the investigation. To do so during the post-accident phase could seriously jeopardize the success of the investigation since it is often essential that the investigator begin examining the circumstances of the accident event as soon as possible. Unnecessary delays could very likely result in the loss of critical evidence and subsequent shortcomings in the investigation. Being prepared saves valuable time and resources, which usually translates into money saved as well.

7

Photography in Accident Investigation

INTRODUCTION

Since the simplification of the photographic process by George Eastman in the late 1880s (Considine 1983), the camera has always been a useful and versatile tool to capture the exact appearance of objects, persons, and scenery. In accident investigation, its utility has proven extremely beneficial. A photograph can accurately record what the human eye may fail to see at the scene of an accident. Critical evidence the investigator may not have noticed when first examining the work environment, specific equipment or materials, debris, or other characteristics of the post-accident scene, may have been captured by the camera and, subsequently, preserved for evaluation at a later date. Photographs also record large amounts of detail and save the investigator a great deal of time and tedious work. The camera has an infallible memory, recording for the investigator minute details that may be needed later during analysis of the evidence. These capabilities can be of immeasurable value to a disciplined and organized investigator.

However, as with any tool used improperly, the camera can actually distort the very evidence it seeks to preserve. Certain characteristics of the camera can be detrimental to a disorganized investigator. For instance, a camera is not as flexible and adaptive to variations in lighting as the human eye, and this adaptation varies even further with the type of film used. Photographic evidence can also be misinterpreted and the result is the same as an eyewitness's lying about what was seen or heard. Errors can be made in exposure, focus, and processing, rendering the photographs useless to the investigator. Serious misunderstanding or misuse of the camera can actually destroy the effectiveness of an investigation.

Perhaps one of the most common errors made by investigators is the tendency to let the camera "think" for the investigator. Taking numerous photographs at random—with blind hope that the key to the accident cause will be captured on film and "discovered" later during the analysis of the photographs—is an approach that not only denies credible analysis of the accident event, but also wastes valuable time as the investigator pours over the massive amounts of unusable photographs in search of evidence that may not exist.

This chapter will introduce the investigator to the benefits and uses of photography and also examine some of its major limitations as an aid to investigation. The intent is to impart a fundamental understanding of the major concepts of basic photography and the techniques that can be applied to the accident investigation process.

PLANNING ACTIVITIES

In the previous chapters of this text, the importance of planning in the investigation process has been repeatedly emphasized. The inclusion of certain types and varieties of camera equipment in the investigation kit, for example, is one aspect of planning that must be considered to ensure effective investigation of most types of accidents.

During planning considerations, one of the first questions to be resolved is whether to utilize the services of a professional photographer during the investigation or to rely on the investigator-manager to take the necessary photographs. The advantages of using professionals are mostly technical in nature; they will typically understand much more about film speed and type, exposure settings, the use of various filters, lighting requirements, and processing, than the investigator. However, experience has shown that, with few exceptions, the disadvantages of using professionals can far outweigh these technical advantages. In most cases, it is uncommon to find a professional photographer who has been trained to see or think like an investigator. The photographer thinks in terms of aesthetic quality and composition of the photographs rather than the analytical use and documentation value of the pictures taken. While the technical aspects of photography will be somewhat easier, the investigator might not be free to work on other tasks while the photographer works to obtain the needed photographs. It is more likely that the investigator will have to accompany the photographer to ensure that those pictures taken reflect an accurate record of the accident scene and to prevent the actions of the photographer from destroying critical evidence in the process. Also, since accidents are generally unannounced events, professional photographers may not always be readily available when needed. Once again, any delays during the post-accident phase could result in the loss of very important evidence.

Because so many variables exist and because proper photography is an essential element of most investigations, it is highly recommended that the investigator-manager be trained in the fundamental skills of photography as an investigative tool. While not particularly complex, the use of camera equipment for investigative purposes will require some familiarization with photography. Aside from the basic data presented in this chapter, the investigator may wish to seek additional information on the subject. Most commercial bookstores or even some hobby shops have a selection of manuals and guidebooks for the amateur photographer. The real enthusiast may wish to enroll in a special course on the principles of photography. Many community colleges offer such courses on a regular basis. Whatever the method used to train the investigator, it is important that it be done during the *planning process* rather than after an accident has occurred. Obtaining an understanding of some basic aspects of photography, such as lighting, filter usage, film speeds, and so on, will certainly facilitate the application of photography as a tool for the investigator. Because many of these basic concepts can be readily applied to accident investigation as well, it should not be difficult to properly assimilate this preparatory training into that of accident investigation.

Planning should also include considerations for initiating photography of the accident scene immediately upon arrival of the responsible investigator or team. Photographs should be taken as soon as possible since evidence is subject to compromise from movement of items by the investigators, theft, weather deterioration, erosion marks, oxidation of fracture surfaces, and so on. Avoiding delays is necessary to obtain good photographic coverage of the accident environment. Consequently, planning of photography is always advisable for effective investigations. It is also recommended that each investigator be fully equipped to make photographs that are commensurate with his or her responsibilities in the investigation. Investigators should know best what it is they are looking for or what is not "correct" about a particular scene or circumstance. Line managers or supervisors should normally be perceptive of the working conditions in their assigned areas. Understanding the value of photographs to document these conditions is an essential part of investigation training. While most accidents investigated solely by an area supervisor may not require extensive photography, the final report would certainly be enhanced by the use of photographs. But when certain material failures may become a pivotal aspect of some future product liability litigation, or when equipment abuse or deficiency in standards can be documented through the use of photographic evidence, then the supervisor should be aware of such potential. Guidelines to supervisors should be published in company accident and incident investigation policy so that supervisors will know when to take or request photographs.

USES OF PHOTOGRAPHY

As a record of evidence, a photograph can accomplish a great deal while saving a tremendous amount of time. The familiar phrase "a picture is worth a thousand words" best describes the extent of the utility a single photograph can have to the accident investigator. Examples of some of the more *general uses* an investigator can make of photography at the accident scene include the following:

1. Photographs can provide an orientation to the scene of the accident in terms of surroundings, direction of travel (if applicable), obstacles, and a general view of the overall environment.
2. Photographs provide an instant record of the detail and extent of injury and damage resulting from the accident event.
3. With just a few properly taken photographs, the investigator can accurately record the exact, post-accident positions of large quantities of items, debris, or damage fragments.
4. Using the camera as a type of "eye" witness, the investigator can obtain a depiction of the scene as viewed by actual witnesses to the accident.
5. Photographs can often be used to document evidence of faulty assembly or improper use of equipment, materials, or structures.
6. Depending upon the nature of the accident, photographs can be used to document the location and extent of any marks, spills, instructional aids, placards, warning signs, and so on.
7. Any evidence of deterioration, degradation, or even abuse and lack of proper maintenance can often be clearly documented using photographs.
8. The location of small parts, or other material evidence, that may have been overlooked during the initial investigation, are often "discovered" during subsequent examination and analysis of photographic evidence.
9. Studying accident scene photographs may make it possible to obtain an clearer understanding of the sequence of machine failure that resulted in the accident event.
10. When completing accident analysis worksheets, fault trees, or other such analytical methods, photographs provide a clear reference and may simplify the overall analysis.
11. Photographs provide clear illustrations for use during follow-on training and implementation of any corrective actions.

In addition to the general examples cited above, photographs can also be used to show *specific elements* of the accident scene that can greatly simplify the investigation process. For example, weather conditions, if outdoors, can change very quickly. Photographs taken immediately after the accident can document the effects of visibility restrictions (caused by rain, snow, fog, haze, smog, smoke, and so on), surface wetting/icing, and other weather factors at the accident site that may or may not have contributed to the loss event being investigated. In fact, photographs documenting weather conditions often prove more valuable than meteorological reports since weather phenomena are highly localized at times.

Photographs taken along the approach path to the accident site, if such movement was part of the pre-contact phase, can often show more than just a view of the scene. Obstructions, such as equipment, debris, and improperly stored materials, as well as other factors like pot holes and improper signs, should be documented as soon as possible after the accident event. Photographs of the general accident scene should always include views from all four sides or major directions. The major directions could be from corners or sides of a room, from intersecting aisle-ways in a warehouse, intersecting roadways, or the major points of the compass (i.e., north, south, east, and west). Properly taken, general views will also greatly facilitate the analysis of other photographs as well as help prevent the misinterpretation of evidence due to factors like camera position, lighting, and shadows. Taking these photographs immediately upon arrival at the scene may later reveal that items were removed or displaced before they were analyzed and recorded.

If possible, the investigator should attempt to obtain overhead views of the accident scene. These not only provide valuable orientation to the scene, they may also help resolve questions that may arise during the examination of other photographic evidence. Such photographs can be obtained from atop surrounding buildings, from stacks of materials, from a high-lift truck, or even from nearby hillsides. While it is important to capture a view of the "big picture" at an accident site, investigators should avoid situations where doing so could cause further injuries, either to themselves or those in the area. Of course climbing unstable structures or standing on unsafe surfaces to make such photographs should not be attempted.

TYPES OF PHOTOGRAPHIC EQUIPMENT

Modern technology offers a wide variety of choice to the investigator wishing to use photography as an investigative tool. However, since certain financial resources are required to invest in camera equipment, the equipment selected will depend first upon a preliminary decision by management and a stated company policy on the nature and extent of the accident investigation effort. Once a decision has been made to develop and implement an accident investigation and loss control program, the investigator can better decide the specifics of the program's photographic support requirements. For example, compact or miniature cameras, Polaroid (or direct positive image) cameras, and video-cameras each have unique uses in the investigation process, as briefly discussed in the following paragraphs. Perhaps the most versatile and therefore effective camera for general investigation is the 35mm single lens reflex camera.

The Compact or Miniature Camera

Often referred to as a pocket camera, the compactness, built-in flash, and easy-load film cassette of the compact camera offer the investigator a low-cost, basic photographic capability. It is portable and requires very little storage space when not in use. This type of camera is most suitable for the general recording and documenting of an accident event. It is relatively easy to use and very suitable to provide line supervisors a means of depicting damages in accidents they must investigate. Another type of camera in this category is the relatively inexpensive "disposable" compact camera, preloaded with film by the manufacturer and sent back to the processor for replacement of both camera and film.

It should be noted that the compact camera is not particularly capable of making the close-up, detailed photographs needed for failure analysis of materials. It is also not the best choice for wide-scene or night photography, for selective focus (to remove background items), or for timed exposures, when required. But, for the general documentation requirements of even the most complex accident investigation, the compact can serve a quite useful purpose.

The Polaroid™

In 1947, E. H. Land demonstrated a "new type" of camera that produced a finished black-and-white positive print, without the need for a negative, available to the photographer within approximately one minute. This was the first model of the *Polaroid* camera (Considine 1983). Today's Polaroid camera, technically advanced and easily adapted to the accident investigation process, is a compact and easy-to-use method of obtaining an instant, full-color, detailed record of a wide variety of desired images at the accident scene.

In general, the Polaroid camera, or any other type capable of producing *direct positive images,* has essentially the same capabilities and limitations as the compact camera described above. But the Polaroid camera and its film supplies are usually larger than the compact cameras and, consequently, take up more of the limited space in the investigator's kit. Unlike most compacts, the Polaroid can be adapted for close-up detail of material and equipment fractures and damage. Aside from acting as a back-up camera during the investigation, perhaps its greatest value is its capacity to provide the investigator with instant documentation during disassembly of equipment. By photographing each successive step in the disassembly process, reconstruction will be greatly facilitated. Also, any notes, instructions, observations, or simple highlighting marks can be recorded almost immediately on the Polaroid print before the next step in the disassembly process is made. Objects of specific interest, captured on the Polaroid picture, can be immediately cross-referenced with other investigation techniques such as sketching and mapping for additional clarity.

Investigators can also use the Polaroid to obtain an immediate reference view of a picture they may wish to take with the compact or the 35mm camera. As a supplementary camera, the Polaroid is excellent for color definition analysis of fire and smoke damage to structures and equipment. Perhaps the single most important advantage of the Polaroid camera in accident investigation is its unique capability of providing the investigator/photographer with relatively clear pictures for immediate confirmation that the picture taken actually shows what was intended before anything at the scene is touched, moved, or removed.

The Video Camera

A video camera will offer the investigator a great deal of utility to record actual movements (of people, machinery), conditions (weather, lighting), sounds (noise or other distractions), positions (of equipment, materials, parts, debris), and other factors relevant to the investigation of an accident event. However, this capability, while convenient and useful, may not always be necessary for every accident situation. The expense of video equipment will usually dictate extensive justification, such as its *multiple-use* capability, before management will agree to the investment: For instance, if the camera can be obtained for training purposes, special documentaries, or other company activities when not being used specifically for investigation, there is usually a better chance of having one available should an accident occur.

Video cameras can be compact, relatively easy to use (in most cases), and quite versatile. Features such as auto-focus, zoom, instant color-recording, automatic light-sensing, quick-load cassettes, portability, extended battery life, and so on, offer the investigator so much more than the still-photo cameras discussed in this chapter. Video cameras with mini-cassettes (usually 8mm or smaller), typically lightweight and quite compact, are even referred to as *palm-corders* because they simply fit into the palm of one's hand.

Unlike the photograph, video records taken at and around the accident scene cannot obviously be included in the final written accident investigation report. But videos still offer the investigator several unique advantages, including their capacity to provide accident site orientation, especially in aircraft or vehicle accident investigation when the approach to the accident site requires analysis. It also has application in the industrial setting as well. It can be used to record a walking or driving approach to, and passage through, an accident site. The film record is useful in reconstruction or in obtaining feedback during any post-accident follow-on training sessions. Videos also allow the investigator to evaluate the actual sequence and timing of events and response actions that occurred during the post-contact phase.

During presentations to upper management, film records taken at the accident scene offer a great deal of clarity that can be easily supplemented with additional still photographs and written descriptions. Videocassette players can be made to pause or freeze the video at any given moment during a presentation to allow for further explanation, emphasis, or description of the material being presented. As with photographs, video records can also be reviewed and analyzed over and over, until each aspect of the record is understood and explained.

Additionally, video records taken during the investigation of past accident events can be quite useful when training the current management staff in the methods and techniques of accident investigation. Actual footage taken in work areas that trainees are familiar with will add a certain level of reality to the investigation-training process.

One potential disadvantage to the use of a video camera in accident investigation is the tendency for some users to forget that the camera will actually record everything while the record button is pressed. We have all seen examples of bad amateur videos that move around too much or too quickly, do not focus properly, or zoom in and out so frequently that it is almost nauseating to watch. A little practice with slow movements, zooming, and gradual focus techniques will go a long way toward the proper use of the video camera during an actual accident investigation.

The 35mm Single Lens Reflex Camera

The 35mm (millimeter) single lens reflex (SLR) camera provides the investigator with the best combination of versatility, portability, and cost for both general and specific use purposes. The 35mm's small film cartridges, or *magazines,* allow the investigator to carry a large supply of various types of film to interchange between color and black and white films of various contrasts, speeds, light sensitivities, and other such characteristics. Most models have built-in light meters that indicate when the lighting conditions will require a supplemental flash. Other characteristics of many models of the 35mm SLR camera, such as easy loading, quick wind and auto-film advance, automatic film rewind, built-in flash (or easily attachable flash unit), tripod mounting capability, fully *automatic* or *programmable* distance and speed settings (on some models), further exemplify the versatility of this type of camera. In addition, multiple interchangeable lenses, from wide-angle to telephoto, give the investigator the capability to photograph any aspect of the accident site precisely as desired. The *through-the-lens* viewing feature of the 35mm camera has several advantages over the range-finder viewing common to other types of cameras. It enables selective focusing, used to exclude extraneous or unwanted details from the photograph (i.e., the undesired objects will appear out of focus in the printed photograph). It also permits the extreme detail of close-up photography, without having to first assemble extra attachments or perform computations of parallax (changes in appearance caused by changes in angular viewing) and depth of focus. Through-the-lens viewing will also allow the photographer to see any light reflections, shadows, and other distortions as they will appear in the photograph. Unlike the compact and the Polaroid cameras discussed earlier, the 35mm camera permits selection of shutter speed to stop movement and control focus. It can be set for timed exposure of night or dimly-lit accident scenes, and most models have a shutter delay or timer feature that enables investigators to photograph themselves holding items or materials and pointing out particular features or positions of evidence.

THE INVESTIGATION KIT

In the discussion about the accident investigation kit in the previous chapter, photographic equipment was listed among *specific* items for the kit (reference Table 6-1). To further facilitate the collection of photographic equipment to be used during the investigation, it is suggested that the investigator keep these items and materials in a separate *photographic package.* This will allow for easy placement into the investigation kit when needed and also will provide an added level of security to protect the high-cost photographic equipment from theft or damage. This package should include at least the following items:

1. *Cameras:* Compact, Polaroid, 35mm, video, or any combination of these photographic capabilities as determined necessary during the investigation planning process.

2. *Auxiliary lenses:* As desired (wide-angle, telephoto, microscopic) for use with the 35mm camera.
3. *Flash unit:* If applicable (i.e., if the 35mm camera requires an attachable unit).
4. *Film supply:* An ample quantity of film cartridges, of varying speeds in black and white as well as color formats, for each type of camera in the kit. Also, a supply of videocassette cartridges if a video camera is to be used.
5. *Lens care materials:* Lens tissues and special cleaning fluid.
6. *Tripod:* For use with the 35mm camera, especially where stability is a problem, or when timed exposures or night photography is required or anticipated. Tripods are also quite effective when using the video camera.
7. *Light blue cloth:* A light blue cloth or towel, 2 feet by 3 feet (or larger), is particularly useful as a background for small items or components being disassembled. Light blue is recommended because it photographs as white on black and white film, allowing the object to stand out in the photograph. Light blue is preferable to white because white cloth reflects light and produces glare. White also causes false light-meter readings that will result in heavy shadows and a subsequent loss of detail in the photograph.
8. *White linen cloth:* A thin, small white cloth or handkerchief can be used to reduce light output of the flash unit when taking close-up pictures, which will allow more detail into the photograph. Too much light at a close-up range will literally flood or wash out desired detail on objects.
9. *Photographic data logbook:* It is essential, for clarification and tracking purposes, for the investigator to keep a detailed log record of each photograph taken to prevent undesired duplication and ensure the later identification of each picture taken at the scene.
10. *Grease marking pencil:* When Polaroid pictures are taken, a grease pencil is used for immediate marking and labeling of the printed photographs at the accident scene.
11. *Scale or ruler:* Useful when photographed with an object to show a relative size reference or to add perspective to the object being photographed.
12. *Pen light:* A small flashlight becomes a very useful tool in dimly lit or nighttime situations when attempting to load film into cameras, set up a tripod mounting, or interchange lenses. Also, when night lighting is insufficient for focusing purposes, the pen light may provide enough illumination to obtain proper focus on the subject being photographed.
13. *Adhesive labels:* To identify used film cartridges, an assortment of adhesive labels is recommended. The photographer can easily cross-

reference labeled film cartridges, with information about each picture also recorded in the photo logbook.

14. *Assorted lens filters:* Filters for screening ultraviolet light, color enhancement, glare reduction, and so on, should be available in the kit for use with the 35mm camera, as conditions warrant.

15. *Perspective grid:* Used as a template to show size differential in a photograph. First developed in 1916 for crime photography, the template is typically constructed of several one-foot-square pieces of floor tile, taped together at their adjoining sides with duct tape. In most cases, the tiles are cut diagonally and then taped back together. This makes folding and storing the grid much easier and also allows the investigating photographer to decide how much of the grid to use for each photograph. The use of the grid is optional and, in fact, may not always be appropriate to the specific scene being photographed. But, when taking a picture of the accident scene, the template can make things easier for later viewing, if relative size and distance between objects can be shown without much analysis. This is the utility of the perspective grid in accident photography.

METHODS AND TECHNIQUES

The ability to capture film images that are useful in accident cause determination depends more on the photographer's approach to the task than on actual talent itself, although the latter would certainly be an advantage in any case. The accident investigator who also acts as the photographer should become thoroughly familiar with the equipment in the photographic package well before a real accident investigation. Familiarization during an investigation will often result in wasted time, effort, and expense, and will seldom produce usable photographic evidence. The investigator should practice using each type of camera, film, and related equipment to determine the specific methods and techniques most suitable to the situations that may arise during an investigation. The photographer should experiment with the use of different film speeds and types, with lighting differences and timed exposures, with the various lenses, filters, and types of cameras (compact, Polaroid, video, 35mm). When practicing, the photographer should experiment with examples of items that may be encountered during an actual investigation. These include vehicles, machines, warehouse aisles, small parts, and general work settings. Sample pictures should be taken indoors as well as outside, in morning, noon, and evening lighting. The familiarization process should be as encompassing as possible.

A basic source of information about photographic techniques is the instruction manual sold with each type of camera. Typically, the manual provides detailed descriptions of every feature of the camera along with helpful hints, tips, and suggestions for optimum use of the equipment. Special techniques and methods can be learned from various guidebooks and other general photography books and manuals available from many local libraries, bookstores, and hobby shops. An understanding of lighting requirements, film and shutter speeds, focusing, and aperture settings is essential for useful photography in accident investigation. Subsequently, these techniques should be second nature to the investigating photographer to ensure maximum benefit during the investigation process. To properly recall the four most important elements of good photography—*shutter speed, aperture, focus,* and *exposure,*—photographers often use the first letters of each, *"S-A-F-E,"* as a natural way to remember.

Using the 35mm SLR Camera

A basic consideration for the investigator using the 35mm SLR camera is *exposure control.* In non-automatic cameras, the user must decide how much light the film should be exposed to, in order to obtain the desired photographic image, and then select the appropriate setting on the camera. On most adjustable cameras, there are essentially two controls that regulate the amount of light reaching the film. These are *shutter speed* and *lens opening* (also known as aperture or *f*/stop). Setting these two controls correctly allows the investigator to take properly exposed pictures. With manual cameras, the user must adjust the shutter speed and aperture controls until the camera's built-in meter indicates that the proper exposure has been set. Automatic cameras will adjust the shutter speed or lens opening (or both) automatically, after determining an optimum exposure setting. Automatic cameras equipped to handle film with DX-encoding designations will set themselves for the speed of the film being used. Because events during the ˙post-accident phase tend to transpire quickly and, because the investigator may not always be able to properly set the 35mm controls (due to inadequate training, lack of time, preoccupation, or otherwise), the automatic version of the 35mm SLR is the preferred camera for investigation purposes. In fact, automatic cameras currently dominate the camera market. Electronic sensors and microprocessors have not only taken the guesswork out of correct exposure, but the labor as well. These cameras will set the shutter speed and aperture (*f*-stop) the moment the shutter release is pressed. Some cameras, which measure the light reflecting off the film itself, can even adjust these settings as the exposure is occurring. However, the relatively higher cost of automatic cameras may be a prohibitive factor in the selection of such a camera.

Whether the camera uses a built-in meter to guide the investigator in setting aperture and shutter speed or sets them itself, the user should still understand the basic premise behind shutter speed and aperture to gain greater control over image quality.

Shutter Speed: The shutter speed controls the length of time the film is exposed to light. Shutter speeds are indicated by the numbers 1, 2, 4, 8, 15, 30, 60, 125, 250, 500, 1000, and 2000. (Some models may not have *all* these numbers indicated on the camera.) Depending upon the sophistication of the camera being used, these speeds may be marked on a dial or shown on a liquid crystal display (LCD) panel on top of the camera or in the viewfinder. The numbers represent *fractions of a second* (except 1 second). For example, the numbers 2, 4, 8, 15, and 30 mean 1/2, 1/4, 1/8, 1/15, and 1/30 of a second, respectively. Choosing the "B" setting on the dial will allow the user to make timed exposures, meaning that the shutter will stay open as long as the shutter release is pressed. These are particularly useful at night when lighting is less than adequate to properly capture an image under normal speed settings. Changing from one shutter speed to another that is nearly twice as fast (for example, 1/60 to 1/125 second) will allow light to strike the film for half as long; therefore, half as much light reaches the film surface. Conversely, changing to a shutter speed that holds the shutter open twice as long (1/60 to 1/30) lets twice as much light to strike the film.

Lens Opening: The size of the lens opening on a camera is the second factor that controls the amount of light that reaches the film. The different sizes of lens openings are indicated by *f*-numbers. These numbers form a series, such as 1.4, 2, 2.8, 5.6, 8, 11, 16, and 22, and are usually marked on the camera lens or shown on an LCD panel. The smallest *f*-number refers to the largest opening; the largest *f*-number is the smallest lens opening. When changing from one lens opening to the next nearest number, the lens has been adjusted by 1 stop. Therefore, moving the setting to the next higher number (for example, *f*/11 to *f*/8) means that the area of the aperture opening is doubled and the film will be exposed to twice as much light. Changing from one lens opening to the next smaller one, *f*/11 to *f*/16, will cut the light in half. Figure 7-1 shows how these numbers will appear on a typical 35mm lens.

Another characteristic of using the 35mm camera, which will require some consideration at an accident scene, is known as *depth of field.* Depth of field is the distance range within which objects in a picture will look sharp or crisp. Investigators who understand this characteristic can use it as a very effective control for making better-quality photographs. Depth of field will vary with the *size* of the lens opening, the *distance* of the subject focused upon, and the *focal length* of the lens. In other words, depth of field becomes *greater* as the size of the lens opening decreases,

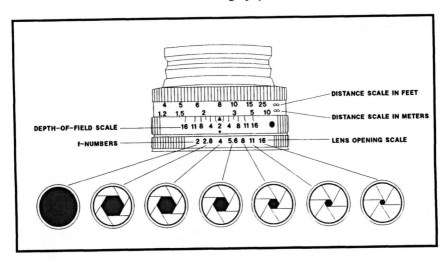

FIGURE 7-1.   The lens openings on a 35mm camera are represented by *f*-numbers, as shown here.

or as the distance to the subject increases or the focal length of the lens decreases and subject distance remains unchanged. Many camera lenses have depth-of-field scales printed on them, as shown in Figure 7-1. Figure 7-2 shows how these variables affect the depth of field. Investigators should understand this characteristic and use it to the advantage of the investigation to capture more usable images as evidence.

The investigator must also choose the speed of the film to be used during the investigation. *Film speed* indicates relative sensitivity to light. The speed is expressed as an ISO (International Standards Organization) number, or as an exposure index (EI), and is usually printed on the box the film came in and on the film magazine itself. The higher the speed, the more sensitive or faster the film; the lower the speed, the less sensitive or slower the film. A fast film requires less light for proper exposure than a slow-speed film. For example, a film with a speed of ISO 25 is slower (i.e., requires more light) than a film with a speed of ISO 200. By setting the film speed on exposure meters and on cameras with built-in meters, the user will obtain the correct exposure for that film. Some cameras automatically set the film speed by reading a special code on DX-encoded films. Table 7-1 provides information concerning the suggested uses of the various types of film speeds for color prints. This information is readily available through photo-processing agencies as well as from the various manufacturers of color films.

For general purposes in accident investigation, film with speeds of 100, 125, 200 or, in most cases, 400 is typically the most useful to the investigator.

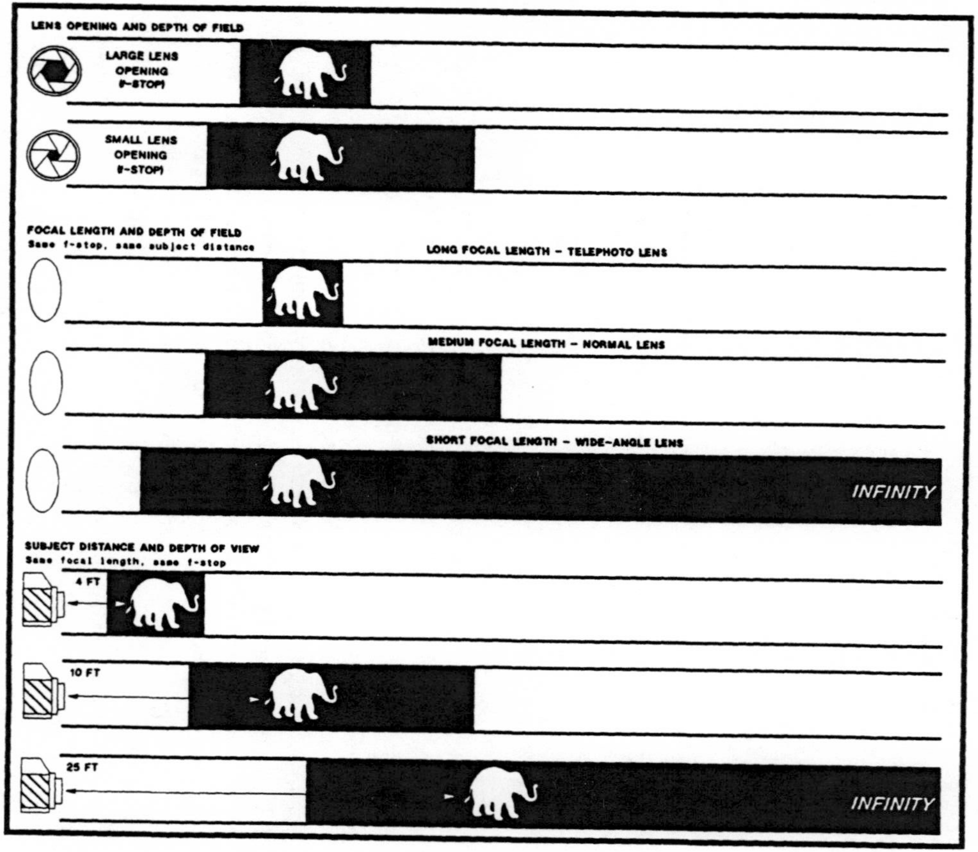

FIGURE 7-2. The concept of depth of field relates to the distance range within which objects in a picture

154

TABLE 7-1   Suggested uses of the various types of film speeds for color prints.

	ISO FILM SPEED	DESCRIPTION	LIGHTING
C O L O R P R I N T S	25	FOR USE WITH SLR CAMERAS ONLY. THE SHARPEST AND FINEST GRAIN COLOR NEGATIVE FILM. USE TO OBTAIN OUTSTANDING, BIG ENLARGEMENTS. EXPOSE IT CAREFULLY SINCE IT HAS LESS LATITUDE THAN THE FASTER-SPEED FILMS.	DAYLIGHT ELECTRONIC FLASH BLUE FLASH
	100	THIS FILM FEATURES VERY HIGH SHARPNESS AND EXTREMELY FINE GRAIN TO ALLOW FOR A HIGH DEGREE OF ENLARGEMENT AND WIDE EXPOSURE LATITUDE. IT IS EXCELLENT FOR USE UNDER GENERAL LIGHTING CONDITIONS.	DAYLIGHT ELECTRONIC FLASH BLUE FLASH
	125	WITH MICRO-FINE GRAIN AND EXTREMELY HIGH SHARPNESS, THIS MEDIUM-SPEED FILM IS AN EXCELLENT CHOICE FOR THE ACCIDENT INVESTIGATION WHERE HIGH QUALITY PRINTS ARE REQUIRED.	DAYLIGHT ELECTRONIC FLASH BLUE FLASH
	200	WITH THE SAME EXTREMELY FINE GRAIN AND NEARLY AS HIGH SHARPNESS AS 100 SPEED FILM, THIS FILM PROVIDES TWICE THE SPEED. IT IS A GOOD CHOICE FOR GENERAL LIGHTING CONDITIONS WHEN HIGHER SHUTTER SPEEDS FOR CAPTURING ACTION OR SMALLER APERTURES FOR INCREASED DEPTH OF FIELD ARE DESIRED.	DAYLIGHT ELECTRONIC FLASH BLUE FLASH
	400	A HIGH-SPEED FILM FOR EXISTING LIGHT SITUATIONS, FAST ACTION GREAT DEPTH OF FIELD, AND EXTENDED FLASH DISTANCE RANGES. MEDIUM SHARPNESS AND EXTREMELY FINE GRAIN OFFER GOOD QUALITY.	EXISTING LIGHT DAYLIGHT ELECTRONIC FLASH BLUE FLASH
	1000	THIS HIGH-SPEED FILM OFFERS VERY FINE GRAIN. USE IT IN DIM LIGHTING WHEN ENLARGEMENTS ARE ANTICIPATED AND THE EXTRA SPEED OF THE ISO 1600 IS NOT REQUIRED.	EXISTING LIGHT DAYLIGHT ELECTRONIC FLASH BLUE FLASH
	1600	VERY HIGH SPEED WITH VERY FINE GRAIN AND MEDIUM SHARPNESS. USE IT IN VERY DIM LIGHTING OR FOR VERY FAST ACTION SUBJECTS.	EXISTING LIGHT DAYLIGHT ELECTRONIC FLASH BLUE FLASH

However, because lighting conditions following an accident event are not very predictable, it is suggested that a small supply of high-speed films (ISO 1000 and 1600) be available to the photographer as well.

Problems Associated with Investigation Photography

There are several inherent characteristics of photography that may affect the appearance of the intended subject(s) in the final printed photograph. These idiosyncrasies are due to the nature of films and camera optics, which to the untrained photographer may appear as defects in the photographic image. In actuality these effects are the result of certain unalterable physical laws that, once understood, can be used to the photographer's advantage. The primary effects result from film color sensitivity, light angle, camera position, lens focal strength, and shadow effects due to film light sensitivity.

Film Color Sensitivity: When black and white film is used, colors of varying hue appear as differing shades of black. White color produces no black, while black color produces complete black. All other colors produce varying shades of partial black, so most appear in print as shades of gray. Each type of film is tailored to produce specific primary colors in desired shades of gray. For example, some films are designed to produce the color blue as a light shade of gray. These films are termed *blue sensitive.* Those which produce red colors as a light gray are *red sensitive,* greens as light gray are *green sensitive,* and so on. General purpose films are typically blue sensitive and will normally record reds and greens as darker shades of gray. This is important to the investigating photographer since it will render blood, hydraulic oils, and other similar stains of potential significance as dark and indistinguishable from oils or from other dark surfaces. To remedy the problem, either corrective filters or color films should be used to show such stains.

As a convenience, a second camera should be available whenever possible to permit the free interchange between color and black and white films. But color films also pose certain peculiarities of which the photographer should be aware. Some color films will reproduce acceptably true colors only in daylight, or only with a flash, or only with supplemental photo-flood incandescent lamps. Within these areas, the films may provide truer color reproduction only in the blue or red end of the light spectrum. This can be a problem when such films are used in areas that are flooded with fluorescent lighting, as is usually the case in most industrial, manufacturing, and shop settings. Consequently, the accident investigator-photographer should experiment with films during the planning and preparation process. Knowing in advance the characteristics of film color sensitivity will allow the photographer to use these effects to the advantage of the investigation.

Light Angle Effects: Lighting is one of the most significant problems in accident photography. When the photographic prints are received, the investigator often finds that vital markings do not show up. Shadows may hide specific details, or light reflections wash them out of the picture. Color sensitivity, discussed above, may have rendered the desired detail indistinguishable from its background because of the angle of light reflection on the photographed surface. Simply stated, the difference between what the investigator expects to see in the photograph and what actually shows up is due to the nature of light and the human eye's ability to adapt to light variations far better than the camera lens and film can. For example, when the object to be photographed has the main light source (the sun, incandescent lamps, or flash lamps) at its back (i.e., *backlighting*), the face of the object will obviously be in shadow and desired detail may not appear as intended on the printed photograph. Flat surfaces, such as roadways, with backlighting will reflect this light into the camera lens and the surface may appear white or washed-out, making identification of important details nearly impossible. To effectively capture surface details on film, *side-lighting* is usually the best option. This is generally true whether the light source is natural or artificial.

Camera Position: Many distortions of the visual picture of the accident scene occur through the characteristic effects of camera position and *angle of view*. The reason, once again, involves the basic differences between the human eye and the camera lens. The investigating photographer should be aware of the simple fact that a pair of human eyes will see objects in stereoscopic or three-dimensional perspective, while the camera lens can only see in two dimensions. The human brain automatically adds mental judgments and comparisons to the images we see with our eyes to produce a mentally perceived picture. A hillside, for example, will always appear to the eye as a slope, regardless of the person's head position (i.e., angle of view). Likewise, a distant object will always be a distant object to the human eye.

The photographic image is quite different. The two dimensions seen by the camera lens provide no other information sources from which to make changes in the perspective views coming through the lens. Therefore, when the camera picture frame is held parallel to the hillside surface, at the same slope angle, the hill will appear flat, while other objects in the frame lean toward the hill, because the sense of gravity is not present in the photograph. Similarly, without reference points for comparison, distant objects will only appear to be small while near objects appear large. (This may or may not have been the case in reality.) This is because the eye and the mind normally see distant objects as *higher up,* toward the horizon, and close objects are seen as low in relation to the horizon. But, if the camera lens is held below normal eye level, the distant object will appear low in the picture and will therefore seem to be a smaller, up-close object, rather than a larger object at a distance.

At an accident scene, photographs should generally be taken from the eye level perspective to ensure an accurate representation of the area. If the lens is held too low, obstructions will appear exaggerated and distant objects will appear too close. Too high a lens angle will reduce the effect of obstructions in the scene and make close objects appear farther away than they actually are.

Focal Length: As shown earlier in Figure 7-2, the focal length of the camera lens will affect the interpretation of distances between objects photographed (depth of field, for example). Focal length can also affect interpretation of the shape of an object. Objects photographed with a lens of short focal length (or, more commonly, a wide-angle lens), will have greater separation between foreground and background than the same objects photographed with a lens of normal focal length. The reverse effect occurs with a telephoto lens; objects appear much closer together and the distance between foreground and background is compressed using a lens of longer focal length.

If the separation of objects is a factor during an accident investigation, the photograph taken with a short focal length, wide-angle lens, could mislead the investigator to erroneously conclude that a greater separation distance existed than was actually the case. In some instances, a round object may appear elliptical and a square object rectangular when photographed close-up with a wide-angle lens. Such photographs should be made from several angles. When photographs of parts are taken for use in examination and analysis, it is consequently important to make note of the lens type in the photo log.

Shadow Effects: Understanding the control and use of shadows is another important aspect of photography in accident investigation. Without shadows in surface depressions, for example, the surfaces will appear flat and depth interpretation and analysis of material fractures will be extremely difficult. The ability to determine the relative height of objects in the photograph will be significantly more difficult with the absence of shadows. Shadow effects in a photograph may also help identify the location of certain objects relative to the accident scene, a fact which may not have been readily apparent during the on-site investigation.

Very simply, shadows can be controlled by the placement of lights in selected locations around the subject or with the use of an automatic flash. Usually, a flash is mounted directly on the camera for user convenience. The light projected forward by the flash at the time the shutter release is pressed will illuminate the face of the subject and cause its shadow to fall behind it, out of the camera's view. Subsequently, very little depth detail will be visible in the face of the subject. With a front-mounted flash, a rough textured surface could actually appear flat and smooth, effectively wiping out any significant details (such as fracture zones) that the photographer intended to show in the photograph. By placing the light source to one side and above the subject to be photographed, the investigator can actually use the shadow effect to the advantage of the investigation to show texture, depth, and surface details in the photographic image. The shadow effects of light placement can be previewed using the flashlight from the investigation kit. When natural light is the only available source, it usually cannot be moved around the subject for best effect. In this case, the photographer must reposition the camera to obtain the best use of the shadow effect.

In short, the photographer-investigator who is observant to shading and light reflections at the accident scene and can adjust the supplemental lighting as well as camera position to obtain the desired images, will make photographs more acceptable for accident analysis and reporting. Those who are not particularly trained or familiar with shading techniques might wish to take several pictures of the same subject using varying degrees of illumination.

PHOTOGRAPHS AS EVIDENCE

As a record of evidence, a properly taken photograph provides the investigator with an extremely detailed and seldom disputable supplement to the written investigation report. Photographs of specific detail become particularly valuable during forensic investigations or in support of testimony given during litigation. As a record of evidence, the photograph can satisfactorily document injury and damage, disassembly of parts and equipment (i.e., reveal evidence of improper assembly, inadequate maintenance, or operator abuse), or show the sequence of failure in a piece of equipment.

Records of Injury and Damage: When injury and damage have occurred as a result of an accident event, the photograph can obviously be used to document the nature and degree of the loss. However, properly taken photographs will also provide the detail necessary to support any subsequent analysis of failure modes and deficient design that may have led to the injury or damage. For instance, the nature of a particular abrasion wound (angle of skin penetration, length, width, depth, and so on) can help investigators identify the exact source of the wound. This could, in turn, lead to recommendations for design changes or the addition of protective measures and modifications to prevent a similar injury from occurring in the future. Although the photographing of the injured and deceased may be extremely unpleasant to most people, including the investigator, it can yield very useful information. Of course, caution and discretion should be exercised to ensure such photographs are not used for purposes other than required to properly conclude the investigative process. The *shock-value* of using particularly graphic photographs as training materials has proven to accomplish relatively little while unnecessarily exposing the viewers to images they probably would rather not see. Also, it should go without saying that the treatment of the injured must never be delayed for the purpose of photographic coverage.

Disassembly of Parts and Equipment: Pictures taken during disassembly of equipment can provide vital evidence of improper assembly or equipment failure due to abuse. Photographs taken in a step-by-step sequence will serve to resolve any subsequent questions that may arise over the proper assembly of the subject equipment.

Photographs of failed, fractured, distorted, or otherwise damaged items are essential to the documentation of analyses and must, therefore, be taken with due care. As each part is identified and tagged, it should also be photographed. Each time a photograph is taken, the subject matter of the picture should be logged, along with its respective photo number, so that proper tracking and accurate identification can occur during any subsequent analysis of each photo. Whenever possible, parts should be photographed using natural light as well as side-lighting or flash to ensure proper recording of all surface details. A scale, ruler, pencil point, or some other object should be used to point out desired details as well as provide a size reference in each photograph.

Sequence of Equipment Failure: In some cases, photographs can be used to demonstrate the sequence of failure that occurred in a piece of equipment that led to the accident event. By photographing the damaged parts in a manner that shows their pre-damaged (or pre-contact phase) positions and then retaking the picture showing the post-contact phase damaged positions, the investigator can literally capture a stop-action sequence of photographs that may reveal or at least help explain how the specific damage could have occurred. These type of photographs can often verify modes of failure, injury, and damage. When the investigating manager can photograph items that are not normally in the work environment (such as mobile equipment, scaffolding, chemicals, stains or pools of liquid), the photographic evidence may help to indicate the timing of the accident event as well as confirm a basic and common accident cause—management's failure to establish proper standards of control in the work area.

Another potential source of photographic evidence, never to be overlooked, is that which may have been taken by the amateur photographer or by the media. Since accidents are usually high drama, newsmedia crews tend to show up rather quickly. In rare instances, their cameras may have captured valuable evidence during the often chaotic moments immediately after an accident event. For example, the color of flames and smoke can reveal clues about the type of materials burned, which may have been totally incinerated by the time the investigators arrived on the scene. Similarly, equipment and materials may have been relocated by fire or rescue personnel before the investigator arrived.

THE PHOTOGRAPH LOG

In most situations, the investigating photographer tends to take a large number of photographs in an effort to ensure the recording of all possible details associated with the accident event. The larger the number of photographs, the greater the potential for loss of intended meaning. When an investigator cannot remember what a particular photograph was supposed to signify or, even worse, cannot even identify the objects in the picture, the chance for erroneous conclusions and assumptions increases and the accuracy of the investigation subsequently suffers.

To avoid such potential, an essential element of accident photography is the *photograph log*. The log is nothing more than a record of the subject matter photographed during the investigation. On the log, each roll of film and each picture on the film are identified, along with the type of equipment used such as lens attachments, filters, lighting methods, angles, and so on. Since each picture is automatically numbered on a roll of film, it is relatively easy to record the appropriate information for each photograph taken by reference to its respective number. The numbering system also facilitates the identification and organization of photographs in the exact sequence they were taken. This is especially important when attempting to show a step-by-step disassembly process. Figure 7-3 is a suggested example of a photograph log, although the exact form used can be tailored to meet the requirements of any investigator.

While photography in accident investigation can be a valuable tool to the investigator, the ability to accurately and consistently identify each picture taken is equally important to the success of the investigation process. The simple and most direct way to ensure the proper identification is to record pertinent information on the photograph log.

SUMMARY

Photography in accident investigation can be a valuable tool for the investigator who is properly trained in the basic uses of photographic equipment. Different types of equipment available to today's investigator include the *compact camera*, the *Polaroid* (or direct positive image) camera, the *35mm single lens reflex (SLR) camera*, and the *video camera*. While having each type available offers the investigator a variety of options, having every type is not entirely necessary to the success of the investigation. The level and degree of photographic capability is a *management decision* that should be resolved during the *planning process* so that the equipment will be readily available if and when an accident occurs. Management may choose to invest in just the compact or Polaroid cameras. They may wish to have a good 35mm SLR camera or even a video camera available for use by their accident investigators. Or they may decide that a combination of all types of equipment will be necessary to properly document and investigate the types of accidents that may occur in their facilities. Whatever the case, these decisions must be made during the planning process, and each manager who might be required to use such equipment should be adequately familiar with

ACCIDENT INVESTIGATION PHOTOGRAPH LOG

FILM IDENTIFICATION NUMBER:			DATE:	
FILM TYPE & SIZE:			ACCIDENT:	

PHOTO #	SUBJECT	LENS	ANGLE	REMARKS
1				
2				
3				
4				
5				
6				
7				
8				
9				
10				
11				

FIGURE 7-3. To ensure accurate reference to photographic evidence, the investigator should keep a photograph log such as that suggested here.

each type *before* an accident investigation becomes necessary. While professional photographers understand many of the technical aspects of photography, they are seldom familiar with the investigation process and, therefore, would require the company of the manager-investigator during photo shoots. Since this saves little time for the investigating manager, it is recommended that organizations train their potential investigators in the proper use of photographic equipment. Also, because accidents generally occur without much advanced warning, it is highly unlikely that a professional photographer could be summoned to the scene at a moment's notice.

Familiarization with basic photographic techniques is essential to ensure the proper use of photography in accident investigation. The user must understand basic concepts such as, but not limited to, *angle of view, positioning, depth of field,* and *lighting effects* before using any camera for investigation purposes. Also, the *investigation kit,* discussed in Chapter 6, should include a *photographic package* containing all the essential items and materials to support the photographic effort during the investigation.

Since the primary objectives of photographic investigation are to gather, document, and preserve evidence for analysis, it is essential that appropriate identification for each photograph taken be recorded on a *photographic log.* Without a method to track and identify the numerous pictures typically taken during an investigation, many will become meaningless and unusable to the investigator.

This chapter provides only a basic understanding of the use of photography during the accident investigation process. It is highly recommended that the investigating manager pursue further independent training on this subject. When used properly, photography can be an indispensable aid to the accident investigator.

8

Collection and Examination of Records

INTRODUCTION

In Chapter 4, the *four P's* or sources of evidence were presented and discussed (reference Figure 4-3). *Paper,* the fourth of the fragile *four P's* of evidence, encompasses all documents, procedures, instructions, manuals, and other types of formal or informal written records that may provide clues as to accident cause when subjected to examination and analysis. In fact, it is not uncommon to find that such records often become the only source of clear evidence that can lead the investigator to identify the basic causal factors of a particular loss event. A thorough analysis of record evidence automatically provides examination of the policies, standards, directives, and specifications that had influence on the working environment at the time the accident occurred. Such records actually forge the attitudes and actions of the people involved in the accident event. Written documentation provides unquestionable testimony that may not be so readily obtainable from an oral or written witness testimony.

This chapter will focus on some specific examples of the types of records that can and should be examined as potential sources of evidence following an accident event. However, as with other aspects of evidence collection, there are no simple guidelines that can be provided to ensure that all types of documentation are properly evaluated. Because each accident event is likely to be different from others, the investigator will initially have to evaluate *all* records that may pertain to a specific accident until knowledge of causal factors can determine which records are relevant to the investigation and which did little to influence the outcome of the event.

SOURCES OF DOCUMENTED EVIDENCE

The concept of *evidence fragility* was introduced and briefly discussed in Chapter 4 (Accident Response Actions). *Paper evidence,* or documents, is considered the least fragile type of potential evidence. It is less likely to be lost, compromised, stolen, damaged, or distorted than evidence that involves *people, parts,* and their *positions* relative to the accident event. Typically, most forms of written evidence are not usually misinterpreted or improperly identified, but they can be misplaced, altered, or overlooked. Also, in many organizations very critical paper evidence, such as operating schedules, work assignments, and the like, may only be maintained for limited time periods (until the end of shift, one week, one month, and so on). Therefore, investigators must make certain that any and all types of potential evidence is collected as soon as possible after the accident event and preserved until such time that each document can be properly evaluated and examined. For instance, in modern industrial settings, the use of computers cannot be overlooked as a possible source of documented evidence. Software errors or inaccurate programming can be the primary cause of an equipment or hardware malfunction that leads directly to the resulting accident event. During any subsequent legal proceedings, it is also important to be able to certify that these documents were collected and preserved unchanged from the time of the accident until they became a matter of record. Also, it is important to examine all types of records in terms of the work situation at the time the record was created and in terms of the situation at the time of the accident. In many cases, written policies fail to keep up with changing technologies and, subsequently, such policies often become antiquated and useless.

To ensure appropriate consideration of this important element of the investigation process, proper planning is once again an essential factor. Planning for accident investigations should include procedures for the identification of any records that would possibly have to be collected as well as a designation of who is responsible for their collection. Since most records are typically not present at the accident scene, they are often forgotten as a source of evidence. Therefore, the time for identifying the likely sources of paper evidence is before the accident occurs. In general, record evidence can be found in the following source categories.

Management Policies

In most cases, activities that result in accidents were (or should have been) under the control of some established management policy or procedure. Recalling the domino effect discussed in Chapter 1 (reference Figure 1-2), *man-*

agement control is the first domino to fall in the accident sequence. Hence, it is also the key element in accident prevention. Therefore, primary sources of potential paper evidence are those records that concern management controls. In most organizations, both general and specific written policies that reflect management's level of interest and control over workplace safety can be found from the executive level down through first-line supervision. It is typically management's policies that cause work conditions to be what they are and actions to be performed or omitted. By examining documented information that describes the expected level of performance of both people and equipment, the investigator can often analyze events to their root cause. Consequently, documents that explain how well performance expectations have been communicated to employees must also be examined (good written policies are worthless if they are not properly communicated to the workforce). Without an appropriate review and analysis of written management polices, the investigation will be less than complete and, possibly, unproductive. The following is a partial listing of some of the sources of evidence typically associated with the establishment of management policy:

1. Recruitment, Selection, and Hiring Records
2. New Employee Indoctrination Documentation
3. Specific Job Instruction Forms
4. Supervisor Selection
5. Management Directives and/or Policy Statements
6. Policies Providing Hazard Protection
7. Personal Protective Equipment Requirements
8. Emergency Response and Preparedness Procedures
9. First-aid Training Records
10. Safety and Loss Control Programs
11. Accident Prevention Programs
12. Historical Records of Accident Events
13. Facility and Equipment Inspection Records
14. Published Rules of Conduct
15. Employee Certification Training Programs/Reports

Many more possibilities exist. Although each company will be different from the next, all should have some level of written policy which must be examined to complete the investigation process.

Risk Assessment and Acceptance

Conditions in a typical working environment, which may be either natural or man-made, can create certain physical as well as psychological stresses on the work force. Other conditions may present a degree of stress on equipment and/or materials. In most cases, the *risk* of encountering such stress may have been *assessed* or evaluated in a form of written analysis and then *accepted* by management. Documents such as work condition reports, hazard analyses, or job safety analyses are all potential records of evidence that may yield specific information regarding the failure of management to identify or properly assess risk. Records that reflect decisions related to risk-hazard analysis can demonstrate management's inability or capability when it comes to risk acceptance. Documentation such as purchase requests, purchase orders, personnel requisitions, and any other that may reflect decisions regarding the allocation of resources, capital expenditures, or the general use of corporate finances may provide valuable information pertaining to the organization's spending policies when it comes to safety in the work environment.

In collecting these records, the investigator must search for any documentation that will demonstrate whether deficiencies were perceived and identified in writing prior to the accident. Then each record must be examined to determine if any substandard conditions were properly evaluated *before* management accepted those conditions. Finally, the investigator must prove whether acceptance of those conditions contributed to or caused the resulting loss event. The following are some examples of the types of record evidence which fall into this particular category:

1. Job Safety (or Job Hazard) Analyses
2. System Safety Analyses
3. Condition Reports and Operation Logs
4. Employee Safety Suggestions (implemented)
5. Employee Safety Suggestions (not implemented)
6. Repair Records Documenting Previous Equipment Failures
7. Risk Assessment Reports

Any other type of documentation that can demonstrate the level and degree of risk assessment and acceptance should be examined. Specific rationale for risk acceptance should also be evaluated to determine if the consequences of such acceptance was properly considered.

Facility and Equipment Design and Maintenance

Paper evidence can also reveal clues about facility and equipment construction and design. Knowing the reasons for a specific design characteristic can often answer questions concerning the work environment that may otherwise go unanswered, or even unasked. Specifications for the design of buildings and equipment, records made during construction or installation, modification decisions made during the design, construction, installation, or use, can all affect the outcome of an accident investigation. For example, the reason for a lack of adequate ventilation, lighting, or other environmental concerns may be found in a record made when the facility was under construction.

The maintenance and care of equipment and facilities is often documented and recorded in preventive or repair maintenance schedules and procedures. Compliance with facility and equipment maintenance standards can be traced through analysis of certain records. Examples of paper evidence which fit into this category include:

1. Building Blueprints and Facility Design Documents
2. Specific Work Area or Section Layout Diagrams
3. Job Specification Drawings and Engineering Orders
4. Construction Progress Charts and Photographs
5. Equipment Installation Manuals and Instructions
6. Parts Lists
7. Maintenance Schedules, Checklists, and Procedures
8. Building Contract Provisions

Facility maintenance records may lead the investigator to insidious causal factors that contributed to the resulting accident event. It is therefore essential that all available records in this category be properly evaluated during the investigation process.

Purchasing Records

Purchasing records were listed, under *Risk Assessment and Acceptance* above, as a method of determining management's policies toward risk assessment and acceptance. In that category, such records that confirm spending policies can, in fact, demonstrate commitment to the safety effort. However, as a separate category, records that document the purchase of equipment and materials can also provide valuable evidence for the investigator. Purchasing requirements can be derived from many sources. The following are only some of the many examples of the paper evidence that can be found in this category:

1. Regulatory Standards and Provisions
2. Proposals to Develop, Fabricate, or Change Parts
3. Management Specifications to Purchasing Department
4. Purchase Orders, Contracts, and Invoices
5. Equipment Purchase Records

Purchasing records may provide documentation of deficiencies in equipment and materials that led to accidents, which can often be traced through paper evidence to inadequate programs, standards and controls.

Personnel and Equipment Certifications

Many job categories require specific skill levels for acceptable personnel performance. Such requirements are normally written and provided as prerequisites for initial or as conditions for continuing employment. Standards for job skill levels can often be found as requirements established by federal, state, and local regulatory agencies. Equipment manufacturers may also stipulate specific skill and operator certification/recertification requirements for personnel designated to operate their equipment. Special uses devised by the company may require additional skills that people need to develop. All these requirements come together in the form of written certification and skill standards. The investigator must ensure that all such documents are carefully reviewed and evaluated. Errors or inadequacies in any of the required skill level assignments or operator certifications might be indicative of serious management shortcomings.

Also, when equipment and machinery are purchased, the manufacturer often provides accompanying documentation that attests that their equipment will operate safely and as expected, provided specific care, maintenance, and operational requirements are met. Investigators should examine the stipulations and provisions of all manufacturers' warranties and compare each with the operational and maintenance history of the subject equipment to verify compliance with manufacturer requirements. Examples of the types of paper evidence common to this category are:

1. Government Regulations
2. Professional, Trade, and Union Standards
3. Company Operating and Maintenance Manuals
4. Personnel Certification Records
5. Manufacturers' Warrantees
6. Material Safety Data Sheets (MSDSs)

Basically, any document that stipulates specific operator skill and certification requirements, as well as those that require certain care and maintenance of equipment, should not be overlooked during an accident investigation. Remembering the discussion of human error in Chapter 5, it is important to note that a causal factor of *human error due to lack of adequate training* is actually a symptom of a deeper problem—management's failure to ensure its employees are properly trained for the jobs they perform.

Public Relations

Where applicable, the investigator should also review records and documents pertaining to any public relations activities that have occurred in or around the accident scene prior to its occurrence. Plant tours and news coverage introduce sources of influence into the work environment that are not normally present during everyday operations. Reviews of plant tour schedules and areas visited might reveal the source of inadvertent (or deliberate) tampering with machine controls and equipment. Retracing the steps of such tours may locate a critical part missing from equipment involved in the accident. News photographs may show significant detail, facts, or conditions about the work environment prior to the accident. The following are some potential sources of paper evidence thath fit into this category:

1. Plant Tour Maps and Schedules
2. Records of Visitor Entry and Exit
3. Visitor Identification Documentation
4. Company Photos Taken During Public Relations Activities
5. News Photographs Taken During Public Access Tours

While not always applicable to every investigation or to every company, documents that record public relations activities should not be overlooked during the investigation. They are yet another potential source of recorded evidence that the investigator cannot afford to ignore.

SUMMARY

This chapter examined *paper evidence* as a source of potential information that should be investigated. With few exceptions, virtually every organization regardless of size will have some degree of documentation on file. Company records and related documents often contain information that may be *crucial* to a thorough investigation of a loss event. Management policies and procedures, records that document spending policies, facility design specifications, personnel training, hazard analysis, risk acceptance, and public relations activities should all be considered during the investigation process.

Any company papers that might be potential sources of evidence must be *identified* in the *planning process* so they are not lost, altered, or overlooked during the investigation. An analysis of paper evidence may be the one link the investigator has, from the *symptoms* of problems to the basic *causes* of accidents.

9

Maps, Sketches, and Drawings

INTRODUCTION

Recalling the *four P's of evidence* discussed in Chapter 4, a major source of evidence for the investigator is the *position* in which parts, people, equipment, and materials are found at the scene of the accident. As a supplement to (and especially in the absence of) the use of photographs, the *position of evidence* can be recorded for subsequent analysis through the use of *sketches, maps, detailed measurements,* and *drawings.* Unlike the photograph, these tools are also quite useful in depicting certain environmental conditions of a particular work area, such as noise-level readings, illumination levels, and concentrations of airborne contaminants. Once such recordings have been made, they can be used by the investigator to facilitate and support the analysis and conclusions provided in the final accident report.

In general, the investigator arriving at an accident should first attempt to *capture* the scene by making a rough sketch or drawing of the area. Detail is not the key at this early stage. The purpose is to record a general view of the accident scene as it appears right after the accident. While photographs also accomplish this and should of course be utilized, the quick sketch or drawing will allow the investigator to make instantaneous notations and comments about the position of evidence right on the drawing. Later, once detailed measurements have been taken, the investigator can use the rough sketches to develop the various types of maps that can then be used as aids during accident analysis.

In some instances, the investigator can also make use of *flow diagrams* to illustrate the sequence or timing of events in the pre-contact, contact, and post-contact stages. These can be particularly useful when analyzing the flow or transfer of energy, work processes, motion, cause and effect relationships in hazard analysis and decision-making, and other such aspects as they relate to the accident event. In the next chapter, System Safety Applications, the use of special diagrams such as *fault trees* will be examined in greater detail. However, for the purpose of general investigation, simple *decision diagrams,* such as that shown as Figure 9-1, can be very useful when attempting to simplify and graphically show the decision-making process.

This chapter will briefly discuss the characteristics and uses of maps, sketches, and drawings and the various aspects of an accident event the investigator must consider to effectively utilize these tools.

PLANNING REQUIREMENTS

Throughout this text, the importance of planning for the overall investigation process has been repeatedly emphasized and discussed. Every aspect of a successful investigation program hinges on proper planning. Mapping is no different in this regard. Planning activities associated with mapping and measurement include surveys of work areas to document normal or routine positioning of fixed items and equipment. A measured sketch showing the location of parts, pieces, and equipment will prove quite useful after an accident occurs. Frequent inspections, with emphasis on the status of these normalities, will keep the management team up-to-date on changes occurring during the normal course of daily business. A side benefit of such a practice is the possibility that the manager may discover abnormal conditions and implement corrective actions before they lead to an accident. A methodical survey conducted as a planning activity will save a considerable amount of valuable time should an accident event occur.

POSITION-MAPPING

The first consideration in the use of mapping during an investigation is the *fragility* of the position of evidence at the scene. The exact position of evidence right after the contact phase of the accident can be very crucial to cause determination. However, the position of parts, equipment, materials, and even people, can be easily moved from the area or relocated to a new position at the scene before or shortly after the investigator arrives. In fact, position evidence is almost always the first form of evidence likely to be altered or tampered with after the accident, for a variety of reasons. For example, injured people must often be moved or removed from the area to receive prompt medical treatment. Fatalities should also be removed from the scene as soon as possible. Damaged or unsafe equipment or structures may have to be stabilized, altered, or removed to reduce the risk of further

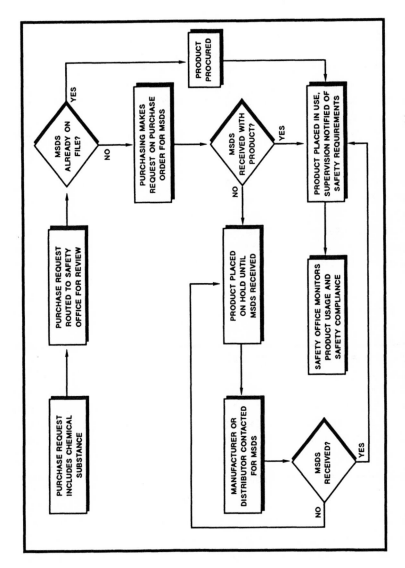

FIGURE 9-1. Example of a typical decision diagram showing the steps in the decision-making process.

loss. High-value items that are also pieces of evidence may be stolen from the accident area, or stored for safekeeping by well-intentioned employees or rescue personnel, and sometimes small parts and items are taken by spectators as souvenirs. Also, in outdoor locations, wind can displace items, rain and snow can obliterate or obscure markings on the ground or alter the size and shape of debris collected at the site. If any alteration of position evidence occurs, the opportunity to capture such evidence is lost forever. At the very least, the result is a prolonged investigation effort. At most, all causal factors may not be properly identified and, subsequently, the investigation effort will fail to accomplish its primary goal of preventing future similar events.

Other problems related to the position of evidence can arise during the analysis if

1. not enough measurements were made to positively establish size and position;
2. the wrong items were measured and recorded;
3. data was not recorded carefully and methodically;
4. photographs alone were relied upon for the recording of position evidence instead of making complete, accurate measurements and sketches.

For these reasons, the training an investigator receives must include ample instructions on the proper use of mapping, measurements, sketches, and drawings as tools to facilitate the investigation. Measurements of distances, dimensions, angles, and weights permit the investigator to formulate precise justifications in support of the accident analysis. Through the accurate use of measurements to record position data, the investigator can avoid later deficiencies in the analysis resulting from imprecise recollections, assumptions, and guesswork. Also, if the reconstruction of a particular accident event should become necessary, maps, sketches, and drawings derived from these carefully taken measurements will certainly facilitate the process.

Measurement and Mapping

One of the very first decisions to be made in the use of measurement at an accident scene is to determine what exactly should be measured relevant to the accident. The suggested approach is first to determine what information might be gained if a certain measurement is taken. For instance, a particular measurement might show the results of the contact phase of the accident. The distance and angle that objects were displaced from the line of motion or propelled from the point of contact will help the investigator determine speed, point of impact, direction of impact, and (depending upon the weight of those displaced objects) the force required to move them. Interference factors such as fixed equipment and other obstructions, materials stored in the area, congestion, and so on, will facilitate an understanding of the events that transpired just after the contact phase of the accident. Measurements can also serve to validate witness testimony and make a stronger case towards causal factor determination.

After the investigator has taken the necessary measurements at the scene of the accident, various types of maps can be created for subsequent examination and analysis. *Position maps,* such as that shown in Figure 9-2, can be as simple or as intricate as the situation warrants. In incidents and minor accidents, maps of basic positions and relationships may be sufficient. When the structural integrity of a building or the ergonomics of equipment is in question, maps and tables of measurement may have significantly more meaning during an analysis. From the position of evidence at the scene, the investigator must determine what happened, how it happened, and why it happened. As stated at the beginning of this section, to accomplish this

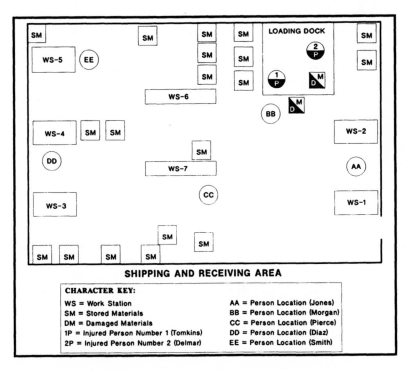

FIGURE 9-2.   Position-mapping documents the basic location of evidence and general appearance of the accident scene immediately after its occurrence.

objective the investigator must first identify those items that must be measured and recorded. Depending on the complexity of an accident event, it may not be entirely possible for the investigator to accurately make such identifications during the often chaotic minutes following an accident. Consequently, it is recommended that the investigator look for and attempt to preserve all possible clues, even if some (or most) of them will be discarded after analysis. Therefore, the location of all items found at the scene of an accident, from the injured and deceased to damaged equipment and materials should not be ruled out as insignificant to the accident until such an assumption can be proven.

Some of the many possible items that should be recorded by position include, but are certainly not limited to, the following:

1. dead and injured people;
2. machinery, equipment, and other items involved in or affected by the accident;
3. parts broken-off of or detached from equipment and materials;
4. objects that were broken, struck, or otherwise damaged as a result of an accident;
5. gouges, scratches, dents, paint smears, rubber skid marks, and so on, discovered on any surface in the accident area;
6. tracks or similar indications of movement;
7. defects or irregularities in surfaces;
8. accumulation stains resulting from liquids that were either present before or spilled during the accident event;
9. spilled or contaminated materials and parts;
10. areas of debris;
11. sources of distraction or adverse environmental conditions such as noise, illumination, vibration as well as weather phenomena;
12. the position of safety devices and protective equipment after the accident occurred.

Although some of the above suggested items (people, noise, illumination, vibration, weather) are not simply "parts," they are still important elements to be considered during the investigation process. For instance, the positions of people who were injured in the accident, who were principals in the accident, or who were witnesses to the accident are vital considerations as evidence to be recorded for subsequent analysis. Details of their positions and movements are necessary to reconstruct their actions and determine the effect, if any, those actions had on the causes of the accident event. Such information is also constructive when evaluating the hazards of systems, processes, procedures, and equipment.

In addition to measuring and mapping those items found at the accident scene, it can be equally important to look for things that should be there but are missing. What is not found at the accident scene can often be essential for cause determination. A critical part to a machine may have become loose or knocked off to some distance from the accident scene. Perhaps it was not replaced after the equipment underwent routine maintenance. Maybe a part that appears on equipment blueprints or in its specifications was never installed by the manufacturer.

Other forms of mapping that can be developed from some of the suggested items listed above is sometimes referred to as *environmental mapping*. Noise, temperature, illumination, vibration, air flow, and other similar environmental factors can have great influence on the conditions that result in accidents. While variations in these factors can sometimes affect equipment operation, they are more likely to impact on human performance in the work environment. For example, an otherwise unexplainable operator position may be due to a sharp change in temperature, vibration, noise, chill draft or similar factor that dissuaded the operator from working in the normal, correct position. A sharp, rapid, or abrupt increase in the noise level when moving through the work area may explain why an injured worker did not respond to an alarm, verbal warning, or aural directive. Environmental mapping can graphically show sources of distraction and discomfort in a work area. Figure 9-3 is an example of a noise-level map and Figure 9-4 shows a temperature gradient map.

There are, of course, a variety of other applications of mapping for investigation purposes. Maps of electrical power distribution, hydraulic systems and fluid-flow dynamics, and air-pressure systems can each serve a specific purpose during investigation and analysis.

Aids to Reconstruction

In simple terms, the mapping of an accident site creates an analytical model effectively reconstructing the accident event in a form that can greatly facilitate the investigation. Maps provide the basis for computing forces of impact, pressures within containers or exerted on people and objects, speeds of movement, and braking of moving parts. In the legal arena, maps can also provide the basis of expert witness testimony in defense against a damage or injury claim, or a basis to advance a claim for losses against a supplier.

Because so much can depend upon accident-mapping, great care must be taken to ensure accuracy in the process. When taking measurements at the accident site and when making sketches and measurement tables, the investigator must therefore be precise throughout or a precision model of

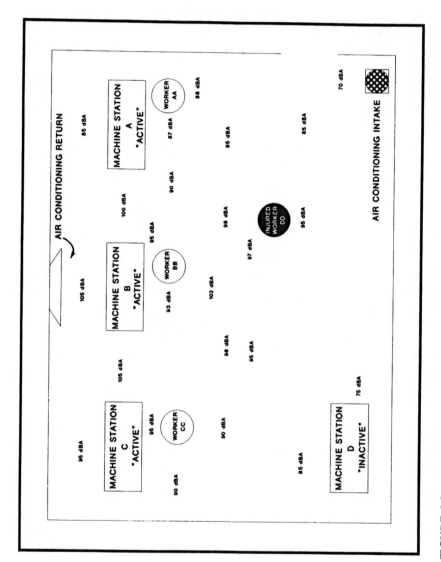

FIGURE 9-3. Environmental mapping of the accident scene, showing noise-level readings taken during the post-contact phase.

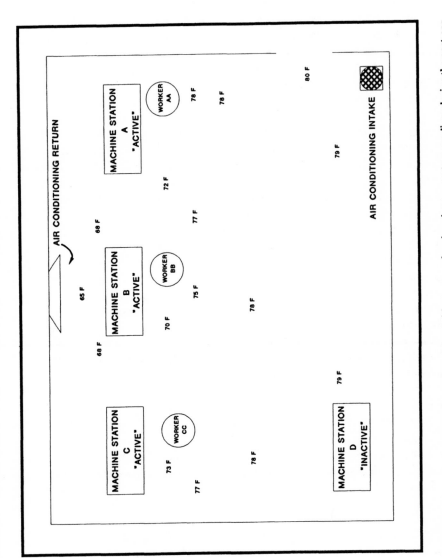

FIGURE 9-4. Environmental mapping of the accident scene, showing air temperature readings during the post-contact phase.

the accident event will not be created. Reconstruction mapping begins at a precise point or object that can be identified later at the site or through the use of photographs, drawings, blueprints, or some other means of verification. Points of measurement should therefore be fixed points, such as numbered posts, signs, columns, poles, aisles, buildings, hydrants, electrical panels, permanently painted markings, machines, fixed furnishings, or other readily identifiable objects. Directional references of north, south, east, and west should also be used on maps, even inside buildings, since these are always constant points of reference.

It is also important to realize that many items to be documented for position data may often require several measurements to accurately fix their position for mapping. In fact, a single measurement will usually be sufficient only when small parts or items are the subject. The exact location of bodies, machines, vehicles, surface marks, and so on, are very important in analysis. The way in which a body, machine, or object fell or moved after contact can be extremely important in reconstruction and analysis as well as subsequent identification of cause. For instance, the position of a body after coming to rest can reveal the nature and intensity of the contact and the forces or actions that occurred during the post-contact phase. Such information also becomes useful during the lessons-learned process, when formulating new plans to prevent injuries and damage in the future. Therefore, the inaccurate placement of critical-position evidence, even if by a few inches, can make the difference in the proper determination of a causal factor. From the legal perspective, such inaccuracies can also lose a case. To further emphasize the importance of proper measuring and mapping, the following are suggested examples of position evidence that should be fixed by a minimum of two measurements:

1. Human Bodies: head, feet, and body lengths (prone position);
2. Machines or Vehicles: two identifiable corners, plus length, width, and height;
3. Gouges and Scars: both ends, plus length; radius of curvatures (if applicable), or deflection of points along mark from a straight reference point;
4. Spots or Debris: outer points or center and diameter, as appropriate, and key or specific items within the debris field.

Clearly, the importance of properly documenting position evidence through mapping cannot be overemphasized. Simply stated, any change that may have occurred between the final position of evidence and its known starting position can reveal the nature of the force of contact (e.g., direction, speed, and so on), which, in turn, will assist the investigator in the ultimate goal of cause determination.

SUMMARY

Accurate *measurement* at the site of an accident is essential to the development of useful sketches and maps. The physical position of evidence is perhaps one of the most critical elements of cause determination. The level of effort required for mapping positions during the actual investigation can be substantially reduced with adequate *planning* before the accident. Preliminary surveys documenting pre-existing conditions and measurements, work area sketches, and other records that show the normal position of fixed objects and equipment, can later be used to develop detailed maps of an accident scene. Because the value of a map to the investigation will rest entirely on its *accuracy*, the importance of *precise measurements* cannot be overemphasized. Also, when reconstruction becomes necessary during the investigation, area maps of the accident site will greatly facilitate the process.

Mapping can be one of the most useful tools an investigator may have to show the effect an accident has had on the work environment. By accurately documenting the position of *people, parts, equipment, machinery, materials,* and so on, both *before* and *after* the accident event, the analysis for cause determination will subsequently be made much easier and more precise.

Maps are also used to depict specific *environmental conditions* that may have had a part in the resulting accident. Factors such as noise, illumination, temperature, an vibration can all affect human as well as equipment performance. A map can plot noise-level readings, temperatures, and so on with graphic detail for the investigator to analyze and evaluate.

While the utilization of mapping techniques is not essential for every investigation, the investigator may find that mapping greatly *facilitates* the overall process and, in most cases, will often lead to a quicker and more accurate conclusion as to the causes of the loss event.

10

System Safety Applications

INTRODUCTION

The use of *system safety applications* in the investigation of accidents was first developed more than three decades ago. The increasing complexity of system designs and the resulting high cost of accidents in terms of human lives and equipment loss created the necessity for a safety assurance approach to the investigation and prevention of loss events. Although system safety methods were originally intended for the evaluation of new systems (during the pre-contact phase), the application of system safety can also be extremely useful tools to the investigator during the post-contact phase of an accident event. It not only provides a way to analyze accidents and incidents after the fact, but also system safety analysis can search for potential system faults *before* an accident ever occurs. The discipline of system safety as an element of systems engineering has its roots in the analysis of safe design for military programs, primarily those dealing with missile and aircraft systems. In fact, it is still used primarily by government agencies such as the *Department of Defense (DOD)* and the *Department of Energy (DOE)*. However, in terms of accident investigation and analysis, the methods developed for and by the military can serve the knowledgeable investigator quite well. Although it has much to offer in terms of accident investigation, certain aspects of system safety are *accident prevention* tools since their main purpose is to identify hazards *before* they lead to a loss.

While the uses of some system safety applications will be briefly discussed in the following pages, the intent is to provide only a simple introduction to the utility of some common system safety applications as tools in accident investigation. The first book in this series, *Basic Guide to System*

Safety, from which much of this chapter has been derived, provides greater detail on the many uses of system safety analysis.

This chapter will focus on the most common and useful system safety applications for accident investigation: primarily, fault tree analysis and management oversight analysis. If additional, more detailed information on this subject is desired, the reader is directed to the *Basic Guide to System Safety.* The investigator should at least be familiar with these concepts, since such analyses can often make the difference between success and failure in accident investigation.

FAULT TREE ANALYSIS

The *Fault Tree Analysis (FTA)* is considered one of the more useful investigative tools in the system safety process, especially when evaluating extremely complex or detailed systems or processes. Because it utilizes the *deductive* method of logic (i.e., moves from the general to the specific), many accident investigators find the FTA very useful in examining the possible conditions that may have lead to the resulting accident. As stated repeatedly in this text, accident events seldom occur as a result of just one initiating factor (i.e., the domino effect). For this reason, in the system safety process of fault tree analysis, the accident event is referred to as the *top event.* This is the *general* or known outcome (i.e., we know the accident occurred) of a possible series of events, the nature of which may or may not be known until investigated. As the investigator begins to identify the *specific* events that contributed to the top event, a fault tree can be constructed. By placing each contributing factor in its respective location on the tree, the investigator can accurately identify where any breakdowns in a system, task, process, or work operation occurred, what relationship exists between events, and what interface occurred (or did not occur, as the case may be).

Although the FTA, by its very name, implies that it is primarily a tool for analyzing faults in a system or process, it is important to note that the FTA can also be used to evaluate the actions necessary to result in *desired* events, such as no accidents. By building a tree depicting all the events that must occur in order to realize the top event of *no accidents,* the manager can use the FTA as a method to construct the foundation of an industrial safety accident prevention program.

Regardless of the intended end result, the use of FTA is a very organized, meticulous, and versatile type of analysis. It is *organized* because it evaluates each event in consideration of that event's specific purpose, function, or place within a system or process. The FTA is *meticulous* because it attempts to describe the relationship of any and all events that may have acted upon a system to result in the top event. This method is also quite *versatile* in its ability to allow for the evaluation of hypothetical events, which the investigator may introduce into the tree to determine potential effects on the top event. In terms of accident prevention, the FTA can also forecast *potential failures* in current design and identify areas where improvement is needed. The investigator can also use the FTA for analyzing normal work activities and operations to determine the nature of real or potential desired or undesired events resulting from system operation. Perhaps one of the more important advantages of the FTA is that it will provide management with more alternatives when evaluating operational safety against operational cost.

Qualitative and Quantitative Reasoning

After all the causal events are listed on a fault tree, the FTA allows the investigator to evaluate each event separately or in combination with other events on the tree. This provides the user with a powerful tool capable of determining, through deduction, which event or set of events led to the top event. When more than one contributing event is identified, as is usually the case, their respective location on the fault tree is referred to as a *cut set*. Identification and qualification of one or multiple cut sets within a fault tree facilitates the evaluation process. Essentially, the cut set isolates specific events in the system and allows for a *qualitative* examination of the relationship between the set, as a whole, and its effect on the top event.

When the likelihood of an event is known and a probability value has been assigned, then analysis of these events on a fault tree will also yield *quantitative* results. As cut sets are identified, the probability of occurrence as a result of cut set interactions can be quantified and the associated risk can be more readily evaluated.

Constructing a Fault Tree

In order to properly construct a fault tree, the investigator (i.e., the manager or supervisor) must first possess extensive knowledge of the system or process under examination. If such knowledge is lacking, then the FTA process must first include in-depth participation from the design community, as well as from other applicable organizational elements within the company (e.g., quality and reliability, operations engineering, facility operations, etc.). The investigator must obtain a clear understanding of the thought process behind the design of the system, as well as any operational criteria that may impact system output.

An understanding of the five elements—people, equipment, material, procedures, and the work environment, as discussed in Chapter 1 (Figure 1-1)—that must all interact for successful business operations, is also essential. Possible causal event factors may exist within any of these elements and must therefore be considered in the FTA process.

The creation of a fault tree begins with the identification of the top event. This event can be as broad and general as *Total System Failure* or as narrow and specific as *Component X Malfunction*. This top event will be placed at the *top of the tree* and all subsequent events that lead to the main event will be placed as *branches* on the tree. Figure 10-1 illustrates the beginning of a

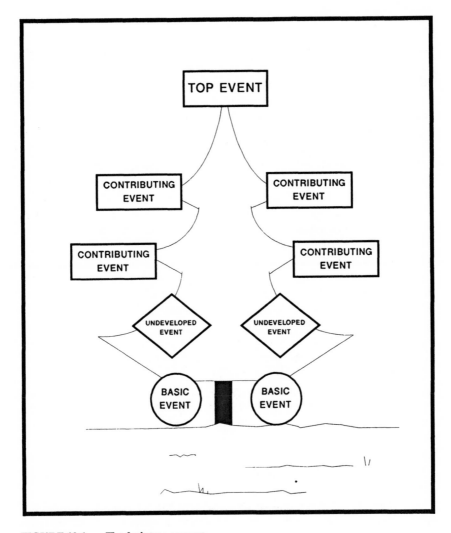

FIGURE 10-1.   The fault tree concept.

simple fault tree, with the location of the top event, the placement of contributing events, and undeveloped events, down to the basic (or *root*) events. As the user moves from the top event downward, each level of the tree will materialize. In order to proceed from one level to the next, the investigator must continually ask the fundamental question: *What could cause this event to occur?* As causal events are identified, they are placed in position on the fault tree.

Fault Tree Symbols

Numerous symbols are used in the construction of a basic fault tree. These symbols, sometimes referred to as *fundamental logic symbols,* provide the user with a pictorial representation of the event and how it interacts with other events on the tree. Figure 10-2 shows the *basic symbols* used during the FTA process. Once the reader has a general understanding of these symbols and their use, as described below, fault tree construction will be greatly facilitated.

The Rectangle: Used to identify the top or primary event (e.g., the accident), as well as secondary or contributing events (sometimes called *main events*). The rectangle shape as used on the fault tree indicates an event, or system state, that must be further analyzed on lower levels within the tree. This is the reason that the primary undesirable event is represented by a rectangle. Everything that appears under it is an attempt to further analyze its occurrence.

The Circle: Used to depict a basic event in the FTA process. It can be a primary fault event (i.e., the first in the process to have occurred) and, therefore, it will require no further development. Use of the circle symbol offers the investigator some flexibility. A *causal chain* could conceivably become quite extensive. Many times, the user will obtain sufficient casual information from analysis of higher-level events in the chain. Therefore, in order not to waste valuable time and resources analyzing a single event to its lowest possible level, the investigator can label a particular event as basic, using the circle symbol indicating that no further development is required.

The House: The house is used to identify a normal event that occurs during system operation. It is an event that either occurs or does not occur, such as turning a switch on or off. It should be noted that, if either the *on* state or the *off* state are possible during normal system operation, then the possible effect of both on the top event should be examined during the investigation.

The Diamond: An event in the fault tree that is considered *undeveloped* is represented by a diamond. Use of this symbol identifies an event that the investigator has chosen not to develop further, either because of the complexity of the event or because insufficient data is available to further analyze the event. Typically, it may also indicate an area of concern where further development should be considered at some point in the future.

The Oval: A *conditional event* or a conditional input into the logic tree that further defines the state of the system that must exist in order for the fault sequence to occur. It may place a restriction on event occurrence based on the occurrence of other events on the causal chain.

Logic Gates: A fault tree is basically a logic tree that shows the association between events on the tree. This association is represented graphically through the use of logic gates. Basically, there are two forms of logic gates that appear on a fault tree: The AND gate and the OR gate (Figure 10-2).

Use of the AND gate means that *all* contributing events connected to the main or primary events through the gate *must* occur in order for the event to occur. For example, when an AND gate is used, if only one or two of three listed events occurs, then the main event *cannot* occur.

The OR gate tells the investigator that, if *either* event connected to a main event through an OR gate occurs, then the main event will also occur. This is an inclusive OR, meaning that occurrence of any or all of the listed events will have the same result. There is an exception to this logic that must be understood. Sometimes, occurrence of one event connected to an OR gate might *exclude* the possibility of other events occurring. For example, in order for a simple electric light bulb to function as intended, it has to be turned on. Events that could lead to failure of the light bulb to activate include filament failure, socket failure, damage to the bulb, faulty wiring, and so on. All these events would be connected to the main event through an OR gate, as shown in Figure 10-3, Analysis A. However, if the scope of the analysis is to consider only one of these possible failures, along with a failure of the light-switching mechanism in the fault tree, the gate is now *conditional* and must be represented by an exclusive OR gate. The oval symbol attached to the gate indicates the condition of the OR gate in this example (Figure 10-3, Analysis B). If more complex systems were evaluated, the exclusive condition placed upon this gate would require further explanation.

In short, the FTA is a technique that can be used by an accident investigator to identify those events that ultimately lead to the undesired outcome. The technique uses a deductive approach to event analysis as it moves from the general to the specific. The FTA has great utility in its ability to distinguish between those events that *must* have occurred (represented by an AND gate) and those that simply *could have* occurred (represented by an OR gate) in order for the top event (the accident) to have occurred. The information charted on a fault tree provides a *qualitative* analysis by dem-

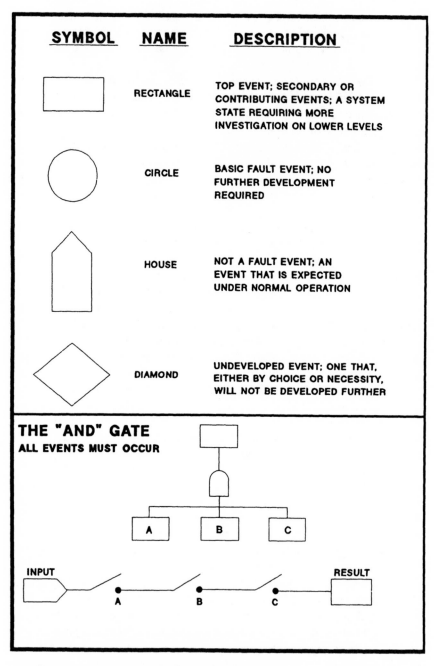

FIGURE 10-2. Standard fault tree analysis (FTA) symbology.

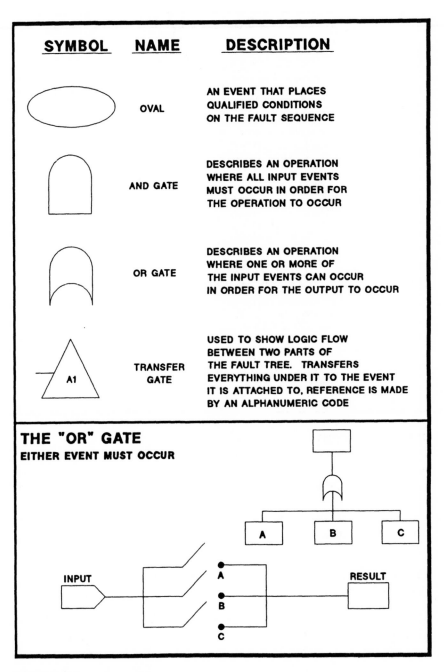

SYMBOL	NAME	DESCRIPTION
OVAL	OVAL	AN EVENT THAT PLACES QUALIFIED CONDITIONS ON THE FAULT SEQUENCE
AND GATE	AND GATE	DESCRIBES AN OPERATION WHERE ALL INPUT EVENTS MUST OCCUR IN ORDER FOR THE OPERATION TO OCCUR
OR GATE	OR GATE	DESCRIBES AN OPERATION WHERE ONE OR MORE OF THE INPUT EVENTS CAN OCCUR IN ORDER FOR THE OUTPUT TO OCCUR
A1	TRANSFER GATE	USED TO SHOW LOGIC FLOW BETWEEN TWO PARTS OF THE FAULT TREE. TRANSFERS EVERYTHING UNDER IT TO THE EVENT IT IS ATTACHED TO, REFERENCE IS MADE BY AN ALPHANUMERIC CODE

THE "OR" GATE
EITHER EVENT MUST OCCUR

A B C

INPUT A B C RESULT

FIGURE 10-2. *(Continued)*

189

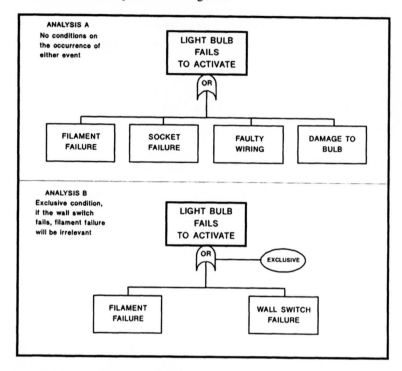

FIGURE 10-3.   The use of exclusive OR gates when proper conditions exist.

onstrating how specific events will effect an outcome. If probability data is known for these events, then the FTA can also provide *quantitative* information to further evaluate the likelihood of achieving the top event.

Once developed, the fault areas that were responsible for yielding the undesired (or desired) event can be evaluated on the *micro* rather than the *macro* level, and this is one of the primary uses of fault tree analysis in accident investigation.

The Management Oversight and Risk Tree (MORT)

Similar in many respects to a fault tree analysis, MORT is essentially a logical analysis of programs, procedures, controls, policies, and all other aspects management can impose to influence the daily functions of the organization. In the investigation of an accident event, MORT systematically examines the adequacy of each of these factors and evaluates their effect on the resulting event. It also has the ability to provide a complete evaluation of a company's entire safety program, even if no accidents have occurred. MORT, therefore, is more than just an accident investigation tool. As with the fault tree, it is also quite effective in the area of accident prevention.

MORT was originally conceived and developed in 1970 by W. G. (Bill) Johnson at the request of the Energy Research and Development Administration (Department of Energy), Division of Safety, Standards, and Compliance. Prior to that time, there was no coordinated and formalized system safety program within the organization. Johnson, who had retired as General Manager of the National Safety Council, advanced the idea that the application of controls and resources by the management of occupational safety and health programs could be categorized into five basic levels (DOE SSDC-4 1983):

1. less-than-minimal compliance with regulations and codes;
2. minimal compliance with regulations and codes;
3. application of manuals and standards;
4. advanced safety programs exemplified by those currently found in leading companies; and,
5. an as-yet-nonexistent, higher-level safety program created by combining the "system safety" concepts pioneered by the military and aerospace industry with the best occupational safety practices and factoring in the newer concepts of behavioral, organizational, and analytical sciences.

Johnson was convinced that a sufficient amount of data already existed to suggest that progression from one level of safety program performance to the next higher (better) level might result in an order-of-magnitude reduction in the annual rate of disastrous accidents experienced by a specific enterprise. His extremely analytical approach to this study yielded the first generation MORT text in 1971. At that time, four key, innovative features, now basic to the MORT program, were introduced:

1. an analytical "logic tree" or diagram, from which MORT derives its name;
2. schematic representation of a dynamic, universal system safety model by using fault tree analysis methodology;
3. methodology for analyzing a specific safety program, through a process of evaluating the adequacy of the implementation of individual safety program elements;
4. a collection of philosophical statements and general advice relative to the application of the MORT system safety concepts and listed criteria by which to make an assessment of the effectively of their application.

The MORT Analytical Chart

Throughout the early- to mid-1970s, the MORT concept was subjected to extensive studies and further development through practical application at test locations. The MORT program, as it exists today, is viewed as a specialized management subsystem that focuses upon programmatic control of industrial safety hazards (DOE SSDC-4 1983). The actual logic diagram displays a structured set of over 1500 basic events, nearly 100 generic *problem areas,* and an unknown number of judging criteria, all interrelated as safety program elements and concepts comprising the ideal safety program model. Figure 10-4 shows the way in which the MORT concept is schematically represented by a logic diagram. The MORT chart has been increasingly used and accepted as a method for analyzing a specific accident or, alternatively, evaluating an existing safety program for accident/incident potential.

As an element of a company's safety management program, MORT is designed to prevent safety-related management oversights, errors, and omissions by providing relatively simple decision-points in a safety program evaluation or an accident analysis. The end results of MORT implementa-

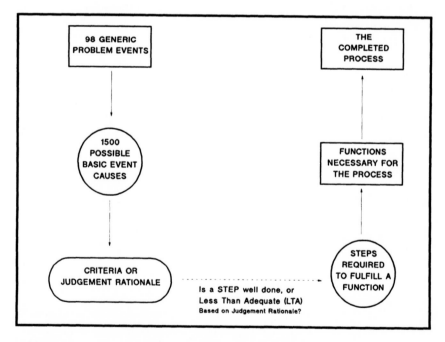

FIGURE 10-4.   The management oversight and risk tree (MORT) process.

tion are the identification, assessment, and referral of residual risks to the proper management levels for appropriate action. MORT will also serve to optimize the allocation of resources available to the safety program and to individual hazard control efforts.

MORT Use

Primarily, MORT has been designed for use as an investigative tool to focus upon the many factors contributing to an incident or accident. It accomplishes this by means of a meticulous tracing of unwanted energy sources, along with consideration of the adequacy of the barriers provided as protection against unwanted contact with those sources (e.g., personal protective equipment, distance/separation, control procedures). As the analysis proceeds, the MORT chart will identify any detected changes in the system (planned or unplanned). When change is detected, MORT recommends the performance of a detailed *change analysis* (DOE SSDC-4 1983).

When system changes are considered, their potential consequences must be evaluated in terms of risk acceptance. Here again, the appropriate management level must determine what is considered *acceptable risk*. Good business management will identify the need for proper control methods, such as barriers, to reduce levels of risk. MORT, then, is designed to investigate accidents and incidents and to evaluate safety programs for potential accident or incident situations. Two of the many basic MORT concepts are the *analysis of change* and the *evaluation of the adequacy of energy barriers* relative to persons or objects (DOE SSDC-4 1983).

The MORT Event Tree

It is not possible within the limited scope of this text to provide fully detailed and comprehensive instruction on the use of the MORT event tree in either safety program evaluation or accident/incident investigation and analysis. Indeed, experts in this technique, such as Bill Johnson, have written numerous textbooks on just this system safety concept alone. The event tree working model, as distributed by the U. S. Department of Energy's System Safety Development Center, is detailed on a single 30" X 24" chart (without instructions). Reproduction of the entire event tree would require several pages alone, which is not practical. Because of the complexity and overwhelming nature of the full MORT event tree, a *mini-MORT* chart has been developed to facilitate the analysis of relatively minor incidents, as well as to serve as a tool for instructor users in the MORT technique. The mini-MORT reduces the number of events to be analyzed and evaluated from 1500 to approximately 150 by eliminating the bottom tier of the chart and removing all transfers from the chart. Since the use of transfer gates, to associate information from one chart to another, is avoided in the mini-MORT, the overall chart tends to appear less complex and intimidating. Additionally, the seldom-used event symbols, like the scroll and the stretched circle, are replaced with the more common circles and rectangles (Stephenson 1991). However, with the mini-MORT, reproduction of the entire event tree would not be possible in a single-page text format. Other excellent alternatives to MORT and mini-MORT, such as Stephenson's *PET chart* (Project Evaluation Tree), also exist but will not be discussed here.

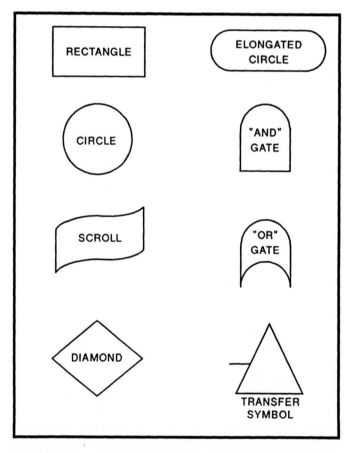

FIGURE 10-5.   MORT symbology. (*Source:* Dept. of Energy *SSDC-4R2 1983.*)

Symbols

The primary symbols used for most analytical trees have been used in the
MORT event tree as well. These include the *rectangle* (primary or top event,
and secondary, contributory or main events), the *diamond* (undeveloped
event), the *circle* (basic event), the AND gate, the OR gate, the *oval*
(conditional or constraint symbol), and the *triangle* (transfer gate or symbol). In addition to these, the MORT chart also uses a *rounded rectangle*, or
elongated circle, to represent a satisfactory event (an event that may have
contributed to an accident or incident but the existence of which is essential
for normal system operation). Also, instead of the *house* symbol common
to fault tree analysis, which represents events that are considered normal
and expected in a typical system, MORT uses a *scroll* symbol. Figure 10-5
shows the MORT symbols as described here.

The MORT analytic diagram works best if it is used as a working paper.
Pertinent facts about an accident or problem may be noted in margins at
appropriate places. Informality is a key factor in the practice of MORT
analysis. The diagram itself will ensure proper discipline in the performance
of the analysis. MORT is a screening guide and a working tool, not the
finished report. (Writing the accident investigation report is a separate
process, described in Chapter 4.) MORT is structured to facilitate analysis
of a given accident event or the potential for its occurrence.

MORT has one strong advantage over most other types of system safety
analyses. It permits the simultaneous analysis of multiple causes through the
tree at any given time. It also isolates specific areas in the sequence of events
where management decision and the assumption of acceptable risk are
required for proper system operation.

SUMMARY

Throughout this text, it has been emphasized that an *organized process* of
analysis is absolutely necessary if the real causes of accidents are to be
determined. In fact, disciplined analysis of accident cause is as time-saving
as it is effective. The *method* an investigator uses to conduct the analysis can
make the difference between a relatively quick and accurate investigation
and one that wanders aimlessly through the facts without developing any
definitive answers.

Accident analysis is a cyclic, repetitive process of examining all available
evidence, forming and validating hypotheses, making assumptions, examining alternatives, and reaching conclusions. A wide variety of methods is
available to accomplish the investigation task in an objective and accurate
manner. This chapter introduced two common *system safety applications,*

which are quite useful during the investigation process: *Fault Tree Analysis (FTA)* and the *Management Oversight and Risk Tree (MORT)*.

The FTA is a *quantitative method* of analyzing the ways and means in which an undesired event, such as an accident, can occur. Through a network of *logic diagrams,* the relationships between the sequence of events, which could have influenced or caused the accident, are analyzed. The FTA does not permit partial or incomplete operation of components to pass as acceptable. On the FTA, anything less than full operation is considered a failure. This characteristic of the FTA prevents speculation and assumption from prematurely dismissing a seemingly insignificant event as a potential cause factor to the accident. With this approach, the investigator can begin to *identify factors* that may have contributed to the overall loss. In other words, all the dominos can be properly identified and evaluated. It is also noted that the FTA can be a useful tool for *accident prevention.* Using the tree concept to determine the causal factors necessary to result in a desired event, such as *no* accidents, the investigator-manager can develop a more complete and effective accident prevention program.

The MORT system of analysis was first developed for the U.S. Energy Research and Development Administration by William G. Johnson. The intent was to provide a safety analysis system that would be compatible with complex, goal-oriented management systems. In his attempt to adapt the concept of systematic analysis and the discipline of the FTA to accident investigation, Johnson devised a complex analytical network of event trees. The resulting chart identified hundreds of immediate and basic causes with their own subordinate areas where lack of adequate management controls may exist.

The purpose of this chapter was to provide the reader with a basic introduction and familiarization with two common system safety analyses that often have useful application in the investigation of accident events. The decision to use system safety in accident investigation is dependent upon many things, not the least of which is the complexity of the system or process to be investigated. While the system safety approach is not always the best way to determine accident cause, it can be effective in most instances. If readers determine that the system safety approach is best suited for their specific circumstances, then further studies on this subject are recommended. Additional, more advanced sources of information and reference are provided in the appendices of this text.

Appendix A

Sources of Additional Information/Training

The following is a compilation of sources where additional, more detailed information and training on the subject of accident investigation and loss control can be obtained. It is noted here that there are certainly many more equally credible references available to the accident investigator than could possibly be listed on these pages. However, those listed here can provide the interested reader of the *Basic Guide to Accident Investigation and Loss Control* with more advanced technical information on accident investigation and prevention.

PROFESSIONAL ORGANIZATIONS

There are numerous national and international professional organizations that either specialize in the practice of accident investigation or have established sections, divisions, or special memberships designed to serve those who investigate accidents. Included among them are the following:

1. American Society of Safety Engineers (ASSE)
   1800 East Oakton Street
   Des Plaines, Illinois 60018-2187

   The ASSE is an international organization with over 26,000 (1993) members internationally. Its membership consists of primarily individual safety professionals dedicated to the advancement of the occupational safety, health, and environmental professions and to fostering the well-being and professional development of its members. Organized in 1911 as the United Association of Casualty Inspectors and incorporated in 1915, the ASSE is one of the oldest sustaining professional safety membership societies in the United States today. Since its inception, with only 35 members in 1911, the ASSE has grown on the fundamental principle of accident prevention and loss control. The monthly publication of the ASSE, *Professional Safety*, often includes articles and information of interest to the loss control professional. Its growing list of technical publications also contains textbooks on the subject of accident investigation. On the national level, the ASSE holds an annual Professional Development Conference (PDC) and quite often offers specialized training and post-conference seminars on accident investigation and loss control. Chapters engage in a number of activities, ranging from monthly meetings with focused presentations to their own specialized seminars and training sessions on a variety of topics. A list of chapters, grouped by states and listed by name and location, can be obtained from the national office.

2. Hazard Control Manager Certification Board
   8009 Carita Court
   Bethesda, Maryland 20817

   Founded in 1976, the Board evaluates and certifies the capabilities of those professionals who are engaged primarily in the administration and management of safety and health programs. Levels of certification are *senior* and *master*, with master being the highest attainable status. The master level indicates that the individual possesses the skill and knowledge necessary to effectively manage comprehensive safety and health programs. The Board offers advice and assistance to those who wish to improve their status in the profession and also works with colleges, universities, and other institutions to establish curricula to better prepare hazard control managers for their duties. It publishes a newsletter entitled *Hazard Control Manager*, which often provides information specifically focused on accident investigation, prevention, and the control of losses.

3. Board of Certified Safety Professionals (BCSP)
   208 Burwash Avenue
   Savoy, Illinois 61874

   The BCSP was originally organized as a peer certification board in 1969 with the purpose of certifying those who practice in the safety profession. The specific functions of the BCSP, as outlined in its charter, are to evaluate the academic and professional experience qualifications of safety professionals, to administer examinations, and issue certifications to those professionals with demonstrated qualifications who have met the BCSP criteria and successfully passed its examinations. It offers examination sessions four times per year (regular spring and fall sessions as well as special offerings at selected professional safety and health conferences during the year). Contact the BCSP for details on the requirements for certification. The Board can also help locate those professionals with expertise in specific disciplines, including accident investigation and loss control.

4. Flight Safety Foundation, Inc. (FSF)
   5510 Columbia Pike
   Arlington, Virginia 22204-3194

The Flight Safety Foundation is an international membership organization dedicated solely to improving the safety of flight. It is nonprofit and independent and serves the public interest by actively supporting and participating in the development and implementation of programs, policies, and procedures affecting safety. To accomplish this, FSF stimulates research into the ways and means of eliminating accident-inducing factors; and it conducts in-depth analysis and appraisals of actual and potential problem areas in flight and ground safety. Once such problems have been identified, solutions are also offered. As a means of information exchange, it also cooperates with all other organizations and individuals in the field of aviation safety in educating all segments of the aviation community in the principles of accident prevention. On that note, the FSF is perhaps most known for its annual safety seminars, held in various parts of the world. These seminars constitute a major gathering of the Foundation's membership and guests, who review and discuss papers presented by prominent safety experts and share ideas for safety improvement in an informal, neutral environment. The FSF also presents awards and otherwise encourages the growth of safety and accident-prevention programs.

5. National Safety Management Society (NSMS)
   3871 Piedmont Avenue
   Oakland, California 94611

The NSMS was founded in 1966 as a nonprofit corporation dedicated to expanding and promoting the role of safety management as an integral element of total management. It accomplishes its objective by developing and perfecting effective methods of improving control of accident losses, be they personnel, property, or financial. Membership is open to those who have management responsibilities related to loss control. The NSMS is an excellent source of information and reference for those managers who are just entering the field of accident investigation as well as for the seasoned professional wishing to expand their knowledge base even further.

6. SAFE Association
   15723 Vanowen Street
   Van Nuys, California 91406

The SAFE Association is a nonprofit professional association dedicated to the preservation of human life. SAFE membership includes representatives from the fields of engineering, psychology, medicine, physiology, management, education, industrial safety, survival training, fire and rescue, law, human factors, and equipment design. It is also associated with the safety and operation of aircraft, automobiles, buses, trucks, trains, spacecraft, and watercraft. Regional chapters sponsor meetings and workshops with specific focus on the various aspects of safety and accident prevention. SAFE publishes a quarterly journal, periodic newsletters, and an annual Proceedings from its SAFE Symposium.

7. System Safety Society (SSS)
   5 Export Drive, Suite A
   Sterling, Virginia 22170-4421

An international not-for-profit organization with membership in the United States as well as 20 countries around the world. It is dedicated to the safety of systems, products, and services through the effective application and implementation of the system safety concept. Originally organized in 1962, the SSS was incorporated in 1973. Its journal, *Hazard Prevention,* is published quarterly and features articles from system safety specialists that focus on the expanding field of system safety, including accident investigation and loss control. The SSS sponsors an International System Safety Conference every other year.

8. Veterans of Safety (VOS)
   203 North Webash Avenue
   Chicago, Illinois 60601

Membership in the VOS is open to those practicing safety professionals with 15 of more years of experience in the safety profession. Founded in 1941, the objective of the VOS is to promote safety in all fields, with specific emphasis on the education of children. Also, through cooperation with the ASSE, it sponsors awards for the best technical papers published during the previous year. In terms of a resource for the accident investigator, the members of the VOS are quite capable of providing advice and information on a variety of related subjects.

9. Human Factors Society (HFS)
   P.O. Box 1369
   Santa Monica, California 90406

   The HFS is a society of psychologists, engineers, physiologists, and other related scientists and professionals who are concerned with the use of human factors in the development of systems and devices of all kinds. *Human Factors,* its official journal, is published bimonthly. Accident investigators will find the HFS an extremely valuable source of reference and information regarding the human element in accident investigation and cause determination.

10. Illuminating Engineering Society of North America (IES)
    345 East 47th Street
    New York, New York 10017

    Although a non-governmental and, therefore, non-regulatory organization, this society is the scientific and engineering stimulus in the field of lighting. It publishes guidelines for proper light illumination in their *IES Lighting Handbook,* the accepted reference for standard lighting requirements. In addition to the Handbook, the IES publishes a monthly magazine, *Lighting Design and Application;* a quarterly *Journal of Illumination Engineering;* and more than 50 other publications on specific lighting concerns in areas such as mining, roadway, office, and emergency lighting. Through the Society and its many publications, the accident investigator can obtain reference material on all aspects of lighting and how the work environment is affected by poor lighting conditions. As a contributing cause to many industrial accidents and incidents, poor lighting is an area that requires serious consideration and understanding during the investigation and analysis process.

11. Industrial Safety Equipment Association (ISEA)
    1901 North Moore Street
    Arlington, Virginia 22209

    The Association has represented the manufacturers of industrial safety equipment since 1934. It is devoted to the promotion of public interest in safety and encourages the development of efficient and practical devices and personal protective equipment for use in industry. The accident investigator can obtain information on codes, standards, and general reference regarding the use of proper protective equipment in the prevention of injury and death resulting from industrial accidents and losses.

12. National Safety Council (NSC)
    1121 Spring Lake Drive
    Itasca, Illinois 60143-3201

    The NSC is dedicated to ensuring the advancement of the safety and health of persons, both on and off the job. Its many excellent seminars, programs, and publications offer the profession quality resources, unmatched on a national level. In addition to a wide variety of activities and resources in the safety and health profession, the NSC offers assistance, training, and printed materials to aid the accident investigator. Its renowned National Safety Congress, which occurs each fall, features a host of effective training sessions and meetings, many of which are of particular interest to the accident investigation and loss control professional.

13. American Insurance Services Group, Inc.
    Engineering and Safety Service
    85 John Street
    New York, New York 10038

    This organization is dedicated to providing information, education, and consultation to the technical and loss control professionals of its participating property-casualty insurance companies. This is accomplished through training programs, conferences, seminars, and a series of reports and bulletins. While the Engineering and Safety Service develops and publishes a wide variety of safety and accident-prevention materials for its subscriber companies, much of this information is also available to the general public. Since the primary objective of insurance is the control of loss, the investigator should find this organization a valuable reference for materials regarding accident prevention.

14. National Fire Protection Association (NFPA)
    Batterymarch Park
    Quincy, Massachusetts 02269

    The NFPA is a clearinghouse for information on the subject of fire protection, fire prevention, and firefighting in general. It is a nonprofit technical and educational organization with a membership consisting of individuals as well as corporations. While most of the technical standards issued by the NFPA will typically be of general interest to the investigator, its *National Electrical Code* (NFPA 70) and *Life Safety Code* (NFPA 101) are of particular interest during most accident investigations.

15. Alliance of American Insurers
1501 Woodfield Road
Schaumburg, Illinois 60195

The Alliance is a national organization of leading property-casualty insurance companies. Its membership is interested in the protection of lives and property by providing protection against mishaps in the workplace and losses from fires, traffic accidents, and other types of incidents. Alliance member companies have a tradition of loss prevention that dates from the organization of the first American mutual insurance company in 1752. Through its Loss Control Department, the Alliance attempts to reduce accidents, fires, and other loss-producing incidents by promoting loss prevention principles and practices. Its many publications, addressing nearly every aspect of accident prevention and loss control, are listed in its catalog, which can be obtained from its New York office. The Alliance works extensively with trade associations and other organizations with similar interests such as the National Safety Council and the American Society of Safety Engineers.

16. System Safety Development Center (SSDC)
U.S. Department of Energy
P.O. Box 1625
Idaho Falls, Idaho 83415

Perhaps the premier authority on the *management oversight and risk tree (MORT)* method of accident or failure analysis, the SSDC has also conducted extensive research and developed informational modules on a wide variety of topics related to accident investigation. These include publications of human factors, change analysis, fault tree analysis, and the development of safety and accident prevention programs. While of specific interest to the system safety professional, the accident investigator will also find these materials an excellent resource.

PUBLICATIONS—BOOKS

In addition to the many publications available through the various organizations and societies listed above, reference books on the subject of accident investigation and loss control include the following suggested titles:

1. Title: *Modern Accident Investigation and Analysis*
Author/Year: Ted S. Ferry/1981 (plus revisions)
Publisher: John Wiley & Sons, Inc.
New York, New York

204 Tools and Techniques for Investigation

2. Title: *Accident Investigation for Supervisors*
   Author/Year: Ted S. Ferry/1993
   Publisher: American Society of Safety Engineers
   Des Plaines, Illinois
3. Title: *Crane Hazards and Their Prevention*
   Author/Year: David V. MacCollum/1993
   Publisher: American Society of Safety Engineers
   Des Plaines, Illinois
4. Title: *Product Safety Management and Engineering*
   Author/Year: Willie Hammer/1993 (2nd Edition)
   Publisher: American Society of Safety Engineers
   Des Plaines, Illinois
5. Title: *Accident Prevention Manual for Industrial Operations*
   Author/Year: National Safety Council/1988 (Ninth Edition)
   Publisher: National Safety Council
   Itasca, Illinois
6. Title: *Damage Control*
   Author/Year: Frank E. Bird and George L. Germain/1976
   Publisher: Institute Press
   Loganville, Georgia
7. Title: *Accident Prevention and Loss Control*
   Author/Year: Charles L. Gilmore/1970
   Publisher: American Management Association
   New York, New York
8. Title: *Professional Accident Investigation: Investigative Methods and Techniques*
   Author/Year: Raymond L. Kuhlman/1977
   Publisher: International Loss Control Institute, Inc.
   Loganville, Georgia
9. Title: *Basic Guide to System Safety*
   Author/Year: Jeffrey W. Vincoli/1993
   Publisher: Van Nostrand Reinhold
   New York, New York

PUBLICATIONS—PERIODICALS

The following is a listing of some of the magazines and periodicals that contain up-to-date information of interest to the accident investigator and loss control specialist. Some are available free of charge to the safety professional, while others are offered for a modest subscription fee. Interested readers should check with the respective publishers listed below. Of course, many more than those magazines listed here exist and, if possible, should also be reviewed on a regular basis. Also, many trade journals for disciplines such as mechanical or electrical engineering often devote sections to industrial safety and accident prevention.

1. Title: *Accident Analysis and Prevention*
   Publisher: Pergamon Press, Inc.
   Fairview Park
   Elmsford, New Jersey 10523
   Frequency: Quarterly
2. Title: *Hazard Prevention*
   Publisher: System Safety Society
   5 Export Drive, Suite A
   Sterling, Virginia 22170-4421
   Frequency: Quarterly
3. Title: *Health and Safety at Work*
   Publisher: McLaren House
   P.O. Box 109, 19 Searbrook Road
   Croydon, Surrey CR9 1QH, England
   Frequency: Monthly
4. Title: *Human Factors*
   Publisher: Human Factors Society, Inc.
   P.O. Box 1369
   Santa Monica, California 90406
   Frequency: Bi-monthly
5. Title: *Journal of Occupational Accidents*
   Publisher: Elsevier Scientific Publishing Company
   52 Vanderbilt Avenue
   New York, New York 10017
   Frequency: Quarterly
6. Title: *Mine Safety and Health*
   Publisher: U.S. Department of Labor
   Mine Safety & Health Administration
   U.S. Government Printing Office
   Washington, D.C. 20402
   Frequency: Bi-monthly
7. Title: *Occupational Hazards*
   Publisher: Penton Publishing, Inc.
   1100 Superior Avenue
   Cleveland, Ohio 44114
   Frequency: Monthly

8. Title: *Occupational Health & Safety*
   Publisher: Medical Publications, Inc.
   225 North New Road
   Waco, Texas 76710
   Frequency: Monthly
9. Title: *Occupational Safety and Health*
   Publisher: Royal Society for the Prevention of Accidents
   Cannon House, The Priory
   Queensway, Birmingham B4 6BS, England
   Frequency: Monthly
10. Title: *Professional Safety*
    Publisher: American Society of Safety Engineers
    1800 East Oakton Street
    Des Plaines, Illinois 60018-2187
    Frequency: Monthly
11. Title: *SAFE Journal*
    Publisher: SAFE Association
    15723 Vanowen Street
    Van Nuys, California 91406
    Frequency: Quarterly
12. Title: *Safety & Health*
    Publisher: National Safety Council
    1121 Spring Lake Drive
    Itasca, Illinois 60143-3201
    Frequency: Monthly

TRAINING SEMINARS/ORGANIZATIONS

There are numerous organizations offering general and specialized training in the area of accident investigation and loss control. In addition to those listed below, many property-casualty insurance companies offer instruction to their clients on the prevention of accidents, the investigation of accidents, and the general control of losses.

1. Sponsor: *Technical Analysis, Inc. (TAI)*
   Address: 2525 Bay Area Boulevard, Suite 200
   Houston, Texas 77058

   TAI is a highly reputable system safety engineering company that offers a wide variety of services to clients in the area of training, engineering consultation, and program administration. Professional training is available in areas ranging from system safety analysis and accident investigation to risk management.

2. Sponsor: *American Society of Safety Engineers (ASSE)*
   Address: 1800 East Oakton Street
   Des Plaines, Illinois 60018-2187

   As described earlier in Appendix A, the American Society of Safety Engineers is an international membership organization of more than 26,000 professionals. The Society sponsors numerous training seminars and a yearly Professional Development Conference. In addition to the many training sessions offered throughout the country, specialized training materials, such as slide/cassette combinations, have been specifically developed to address the accident investigation and loss control arena.

3. Sponsor: *National Safety Council (NSC)*
   Address: 1121 Spring Lake Drive
   Itasca, Illinois 60143-3201

   The NSC is most known for the high-quality training and professional development courses it makes available throughout the year. Through its network of local or regional chapters established nationwide, excellent training opportunities offered by the NSC are accessible to virtually every practicing safety professional in the United States.

4. Sponsor: *COMCO, Inc. Environmental & Safety Services*
   Address: 17121 Clark Avenue
   Bellflower, California 90706-5730

   COMCO is a full-service safety and environmental consulting firm offering specialized training in most aspects of the safety and environmental professions. COMCO is capable of helping its clients with all their safety training needs, including accident investigation and loss control. Its competent staff can also assist clients in the development of specific accident investigation and accident prevention programs.

5. Sponsor: *Summit Training Source (STS), Inc.*
   Address: 6504 28th Street, S.E.
   Grand Rapids, Michigan 49506

   STS develops quality training packages (films, workbooks, study materials, etc.) that can then be purchased to complement the specific training requirements of any given organization. Its materials cover the broad spectrum of safety and health issues, including a film/workbook combination entitled *Accident Investigation in the Workplace*. This film covers the basics of the investigation process and is highly recommended as an introduction to accident investigation.

Appendix B

Acronyms and Abbreviations

In almost every profession or specialty there seems to be a proliferation of the use of acronyms and abbreviations. The profession of accident investigation and loss control is no different. The following is a reference listing of acronyms most frequently used or encountered, either in this text or during the general practice of accident investigation.

ACGIH American Conference of Governmental Industrial Hygienists
AIHA American Industrial Hygiene Association
AMA American Medical Association
ANSI American National Standards Institute
ASSE American Society of Safety Engineers
ASME American Society of Mechanical Engineers
ASTM American Society of Testing Materials

BCSP Board of Certified Safety Professionals

CBA Cost-Benefit Analysis

CGA Compressed Gas Association
CEA Cause and Effect Analysis
CFC Chlorofluorocarbon
CFR Code of Federal Regulations
CPSC Consumer Product Safety Commission
CPR Cardiopulmonary Resuscitation
CSP Certified Safety Professional

DOD Department of Defense
DOE Department of Energy
DOJ Department of Justice

EAP Employee Assistance Program
EI Exposure Index

EPA Environmental Protection Agency
ETBA Energy Trace and Barrier Analysis

FHA Fault Hazard Analysis
FM Factory Mutual
FMEA Failure Mode and Effect Analysis
FR Federal Register
FTA Fault Tree Analysis
FSF Flight Safety Foundation

GIDEP Government-Industry Data Exchange Program

HA Hazard Analysis
HFS Human Factors Society
HTI Hand Tools Institute

IEEE Institute of Electrical and Electronics Engineering
ISO International Standards Organization
IES Illuminating Engineering Society
ISEA International Safety Equipment Association

JSA Job Safety Analysis
JHA Job Hazard Analysis

LCD Liquid Crystal Display
LSC Life Safety Code

MIL-STD Military Standard
MORT Management Oversight and Risk Tree

NASHP National Association of Safety and Health Professionals
NBS National Bureau of Standards

NEC National Electrical Code
NEMA National Electrical Manufacturers Association
NFPA National Fire Protection Association
NSMS National Safety Management Society
NIOSH National Institute for Occupational Safety and Health
NSC National Safety Council
NTSB National Transportation Safety Board

OHA Operating Hazard Analysis
O&SHA Operating and Support Hazard Analysis
OSHA Occupational Safety and Health Administration

PDC Professional Development Conference
PEL Permissible Exposure Limit
PET Project Evaluation Tree
POC Point of Contact
PHA Preliminary Hazard Analysis
PHL Preliminary Hazard List
PPE Personal Protective Equipment
PTI Power Tool Institute
PTSD Post-Traumatic Stress Disorder

SIC Standard Industry Classification
SHA System Hazard Analysis
SLR Single-Lens Reflex
SOP Standard Operating Procedure
SPF Single Point Failure

SSS System Safety Society
SSDC System Safety
    Development Center

TAI Technical Analysis, Inc.
THERP Technique for Human
    Error Rate Prediction
TLV Threshold Limit Value
TQM Total Quality Management

TQMS Total Quality
    Management System
UL Underwriter's Laboratories

VOS Veterans of Safety

WHO World Health
    Organization
WSO World Safety Organization

Appendix C

Accident Investigation Checklist

The success of any accident investigation is almost entirely contingent upon the type of questions asked by the investigator. In fact, the whole investigation process is based on a series of question-and-answer exercises that lead investigators to formulate and test their hypotheses and assumptions, which are later used as a basis for their conclusions. Perhaps one of the most critical characteristics of successful accident investigation is the *quality*, not the quantity, of the questions asked by the investigator.

Appendix C contains an example of an *accident investigation checklist* that can be used to facilitate the investigation process. The critical questions that should be asked are organized and separated into specific categories. Those that apply to a given accident can then be asked and those that do not can be ignored. However, a degree of caution must be taken whenever checklists of any kind are used, especially during the investigation of an accident event. While checklists can be a great tool to assist a manager in organizing the investigation, the user should realize that many more possible questions than those on any generic checklist may have to be asked to understand the specifics of a particular loss event. The investigator should, therefore, use the checklist only as a foundation or starting point to the overall investigation. The checklist should only be used as a means to a start, and not a means to end. In support of an investigation, stopping the pursuit of information upon completion of a checklist is extremely unwise and, in most cases, could result in an incomplete understanding of the accident and *all* its causal factors.

ACCIDENT INVESTIGATION CHECKLIST

CATEGORY I: MANAGEMENT

<u>YES</u>	<u>NO</u>	<u>N/A</u>	<u>QUESTION</u>	<u>REMARKS</u>
___	___	___	Did supervision fail to detect, anticipate, or report an unsafe or hazardous condition?	
___	___	___	Did supervision fail to recognize deviations from the normal job procedure?	
___	___	___	Did the supervisor and employees participate in job review sessions, especially for those jobs performed on an infrequent basis?	
___	___	___	Was supervision made aware of their responsibilities for the safety of their work areas and employees?	
___	___	___	Was supervision properly trained in the principles of accident prevention?	
___	___	___	Did supervision fail to initiate the corrective actions for a known hazardous condition which contributed to this accident event?	
___	___	___	Was there any history of personnel problems or any conflicts with or between supervisor and employees or between employees themselves?	
___	___	___	Did the supervisor conduct regular safety meetings with his or her employees with recorded minutes of the topics discussed and the actions taken?	
___	___	___	Were the proper resources (e.g., equipment, tools, materials, etc.) required to perform the job or task readily available and in proper condition?	
___	___	___	Did supervision ensure employees were trained and proficient before assigning them to their jobs?	
___	___	___	Did management properly research the background and experience level of employees before extending an offer of employment?	

CATEGORY II: EMPLOYEES				
YES	**NO**	**N/A**	**QUESTION**	**REMARKS**
___	___	___	Did a written or well-established procedure exist for employees to follow?	
___	___	___	Did job procedures or standards properly identify the potential hazards of job performance?	
___	___	___	Were employees familiar with job procedures?	
___	___	___	Was there any deviation from the established job procedures?	
___	___	___	Did any mental or physical conditions prevent the employee(s) from properly performing their jobs?	
___	___	___	Were there any tasks in the job considered more demanding or difficult than usual (e.g., strenuous activities, excessive concentration required, etc.)?	
___	___	___	Was the proper personal protective equipment specified for the job or task?	
___	___	___	Were employees trained in the proper use of any personal protective equipment?	
___	___	___	Did the employees use the prescribed personal protective equipment?	
___	___	___	Were employees trained and familiar with the proper emergency procedures, including the use of any special emergency equipment?	
___	___	___	Was there any indication of misuse or abuse of equipment and/or materials at the accident site?	
___	___	___	Is there any history or record of misconduct or poor performance for any employee involved in this accident?	
___	___	___	If applicable, are all employee certifications and training records current and up-to-date?	
___	___	___	Was there any shortage of personnel the day of the accident?	

CATEGORY III: EQUIPMENT				
YES	**NO**	**N/A**	**QUESTION**	**REMARKS**
___	___	___	Were there any defects in equipment (including materials and tools) that contributed to a hazard or created an unsafe condition?	
___	___	___	Were the hazardous or unsafe conditions recognized by management, employees, or both?	
___	___	___	Were the recognized hazardous conditions properly reported?	
___	___	___	Are existing equipment inspection procedures adequately detecting hazardous or unsafe conditions?	
___	___	___	Were the proper equipment and tools being used for the job?	
___	___	___	Were the correct/prescribed tools and equipment readily available at the job site?	
___	___	___	Did employees know how to obtain the proper equipment and tools?	
___	___	___	Did equipment design contribute to operator error?	
___	___	___	Was knowledge of the location/use of emergency equipment required for this job?	
___	___	___	Was all necessary emergency equipment readily available?	
___	___	___	Did emergency equipment function properly?	
___	___	___	Are there any records of system safety analyses for the equipment involved in the accident?	
___	___	___	Is there any history of equipment failure for the same or similar reasons?	
___	___	___	Has the manufacturer issued warnings, *Safe-Alerts,* or other such information pertaining to this equipment?	
___	___	___	Were all equipment guards and warnings functioning properly at the time of the accident?	

CATEGORY IV: WORK ENVIRONMENT

YES	NO	N/A	QUESTION	REMARKS
___	___	___	Did the location of the employees, equipment, and/or materials contribute to the accident?	
___	___	___	Were there any hazardous environmental conditions that may have contributed to the accident?	
___	___	___	Were the hazardous environmental conditions in the work area recognized by employees or supervision?	
___	___	___	Were any actions taken by employees, supervision, or both to eliminate or control environmental hazards?	
___	___	___	Were employees trained to deal with any hazardous environmental conditions that could arise?	
___	___	___	Were employees not assigned to a work area present at the time of the accident event?	
___	___	___	Was sufficient space provided to accomplish the job?	
___	___	___	Was there adequate lighting to properly perform all the assigned tasks associated with the job?	
___	___	___	Did unacceptable noise levels exist at the time of the accident event?	
___	___	___	Was there any known leak of hazardous materials such as radiation, chemicals, or air contaminants?	
___	___	___	Were there any physical environmental hazards, such as excessive vibration, temperature extremes, inadequate air circulation, or ventilation problems?	
___	___	___	If applicable, were there any hazardous environmental conditions, such as inclement weather, that may have contributed to the accident?	
___	___	___	Is the layout of the work area sufficient to preclude or minimize the possibility of distractions from passersby or from other workers in the area?	
___	___	___	Is there a history of environmental problems in this area?	

Appendix D

Case Studies in Accident Investigation

The nature and severity of a given accident event will typically dictate the extent of the investigation required. An investigation of a *minor accident* may require only a minimal investigation effort conducted by the local supervisor with consult from the safety department (see Chapter 3, Figure 3-1); whereas, the effort required to investigate a *catastrophic accident* event will generally require participation from numerous members of management representing a wide variety of specialties. In some cases, outside experts and consultants may have to be brought in to assist the investigation team with specialized examination of accident data (see Chapter 3, Figure 3-2).

Appendix D summarizes the investigation of two different accidents, one relatively minor and the other a catastrophic loss. It should be noted that these events are *true cases* that have occurred in industry. Names, dates, locations, and any other such specific evidence have been changed to protect privacy. It is also noted that the information provided here is presented in summary fashion. The documentation subsequent to the actual investigations fill many volumes and could not be presented here.

As each summary is studied, the reader should be evaluating the information presented in terms of *potential* management oversights and omissions which set the sequence of failure into motion (the domino effect) leading to the loss event. Readers should also consider potential factors which *may not* be clearly presented in each summary. The intention is not for the reader to simply study examples of accident investigations, but to stimulate the thought process as well. It is hoped that information missing from the case studies will foster the investigative intuition of the reader and that the reader will begin to ask questions which go beyond the scope of the information provided. If this is accomplished, then the reader will realize the true nature of accident investigation: Never stop asking questions, even when it appears there is no other information to be discovered.

CASE STUDY 1
Minor Accident Event

Date of Accident: May 18, 1992 (Monday)

Time of Accident: 10:17 a.m.

Location of Accident: Chemical Processing Department
                          Building 601, Room 103A

Supervisor: James Vincent

Losses Incurred: Minor injury
                       Fractured skull; concussion
                       Minor property damage
                       Parts basket dented
                       Metal parts ruined from acid exposure

Summary Description of Event

At 08:05 a.m. on Monday, 18 May 1992, Technician John Doe arrived at his assigned station in Building 601, Room 103A (Chemical Processing). Immediately, he discovered several hundred metal parts already waiting for him. These parts were to be loaded into a holding basket, raised by bridge crane over and into the 2000-gallon acid-etching tank, and then dipped into the next tank that contained water. After rinsing in the water, the parts were to be lifted and lowered into an air-drying bin from which they would later be removed and delivered to the manufacturing department for assembly. Since transferring to this department from the company's shipping department one month ago, John had already received on-the-job training in each phase of the chemical etching process. Although the tremendous amount of parts seemed a bit irregular for a Monday morning, John felt he could handle the task without much difficulty. Department supervisor Jim Vincent was attending a scheduled Monday morning staff meeting with his superiors and the other two employees assigned to this area had not yet arrived. John preceded with the loading of the parts into the specially made metal basket. Because there were so many, it took until approximately 09:45 a.m. for him to complete the loading by himself. Once loaded, John maneuvered the bridge crane over the parts and engaged the automatic hook to the lifting beam attached to the parts basket. Using the crane's remote control pendant, John positioned the parts basket directly over the 2000-gallon acid tank. Because the tank was 8 feet high, John stopped the crane and walked up the catwalk near the tank to verify the positioning of the basket. Once certain it was properly aligned, he lowered the basket into the tank and set the timer. The written procedure for this operation requires the parts to remain in the acid-etch solution for not more than 20 minutes. Since it was just past 10:00 a.m., John decided this would be a great time to take his morning break. Before leaving the area, John disengaged the automatic hook and raised the crane up and away from the acid-gas environment. Shortly after he left for his break, another large load of parts arrived for processing.

At approximately 10:10 a.m., Tom Porter and Jane Ode arrived in the work area. They had been in a scheduled training session for most of the morning. At approximately the same time, supervisor Jim Vincent returned from his staff meeting where the subject of potential layoffs and management cutbacks were the highlight of the meeting. Noticing the parts waiting for processing, he ordered Tom and Jane to start on them immediately. He was visibly disturbed and made mention that his new employee, John Doe, should have already taken care of those parts when he arrived that morning. Jane and Tom loaded the parts into a loading basket and maneuvered the crane over the basket. Once the automatic hook was engaged, they raised the basket over the acid tank and began lowering it into the etching solution. Just as the basket hit the one already in the tank, John Doe returned from break and yelled to stop the crane. The second basket of parts became disengaged from the crane. As it began to roll into the adjacent water tank, the damaged basket came open and several metal parts fell out of the basket; one hit Jane on the head causing her to collapse. Jim Vincent hit the emergency stop switch near his desk, which cut all power to the crane. Jane received a fractured skull and a concussion. The parts in the acid were ruined due to excessive exposure. Both parts baskets were damaged, but repairable.

Investigation Findings

In this particular case, Jim Vincent did not conduct the investigation since he was actually considered part of the cause. His immediate supervisor, Stan Smith, performed the investigation with assistance from the company safety representative. The actual investigation of this case identified numerous shortcomings and potential causal factors. Listed here in no particular order are those that are considered to have been the more serious contributors to the cause of this accident:

1. *Poor Communication in Department:* John Doe was unaware of the reasons his co-workers were not present that morning. He was not informed of the reasons for the excessive parts on a Monday. John Doe left no word of his actions before going on break. James Vincent failed to provide instructions to John Doe and simply assumed the employee would know what to do.

2. *Inadequate Written Procedures:* It became apparent that the written procedure for this job did not contain sufficient warning steps to ensure the safe operation of the crane, the acid-etch steps, or the securing of the parts basket.

3. *Inadequate Safeguards:* The crane did not have any type of lockout mechanism to prevent operation while parts were already in the tank. There was no barrier to prevent employees from entering into the area immediately in front of the tank where Jane Ode was injured.

4. *Training:* John Doe was a relatively new employee to this work area with no real prior experience in the techniques, hazards, and procedures associated with chemical processing. His only real training up to this point was received on the job.

5. *Preoccupation of Supervision:* Jim Vincent was visibly distressed over the topics discussed at his morning staff meeting. He was preoccupied with the issue of layoffs and apparently did not notice the parts

already in the tank. Instead, he mistook the second load of parts as the first and ordered Jane and Tom to process them immediately.

6. *Preoccupation of Employees:* Concerned over their supervisor's obvious mood problem, Jane and Tom did not check the status of the acid tank before proceeding with their orders. John Doe was thinking more about going on his break than he was about securing the work area.

There are numerous additional questions that must be asked in order to identify all potential causal factors in this case. For example:

Why were there so many parts on a Monday morning? Was there a problem with the equipment over the weekend that was not reported?

Why was a relatively new employee left alone to work a hazardous procedure?

Have there been any previous, unreported incidents of this type?

What caused the crane to release the load? (Was it simply because of the impact with the other basket, or was it due to a defective hook mechanism?)

Is there a requirement for personal protective equipment in this area, such as hard hats (when crane is operating)? If not, why not? If so, why was Jane not using it?

This question-and-answer process should continue until all possible causal factors have been identified, considered and evaluated. The checklist provided in Appendix C would be helpful during this stage of the investigation. The recommendations to preclude future similar events may include the installation of lock-out devices on the cranes; the use of warning devices (lights, audible alarms) when processing is attempted while parts are already in the tank; better training for employees and supervisors, adequate procedures to address *all* the hazards of the operation; and an enforced policy for the use of personal protective equipment. There are certainly many other possible corrective actions that can prevent recurrence. Readers are invited to further explore these possibilities on their own. Consider those questions already asked, the answers provided, and other questions that have not yet been asked. Although the answers may not be verified through the information provided here, the point of the exercise is to experience the analytical aspect of investigation. In other words, ask, analyze, evaluate, and consider all aspects.

CASE STUDY 2
Catastrophic Accident Event

Date of Accident:	July 15, 1991 (Monday)
Time of Accident:	09:40 a.m.
Location of Accident:	Underground Fuel Vault Area 57, Fuel Farm #1
Supervisor:	Ted Whitt
Losses Incurred:	Death, three employees

Summary Description of Event

At XYZ Refinery's Area 57 there are three fuel farms that are used to store a variety of bulk fuels including gasoline, kerosene, and diesel fuel. Fuel farm #1 has an underground vault of specific dimensions 15 feet high, 18 feet wide, and 50 feet long. The vault is constructed of masonry/concrete walls and ceiling with an earthen/gravel/soil floor surface. Inside the vault there are several steel tanks that contain or have previously contained kerosene fuel. The fuel is dispensed from the underground tanks into waiting trucks through surface-mounted transfer pumps.

In the summer 1991, XYZ Refinery began receiving complaints from customers who purchased kerosene fuel. By the second week of July, thirteen complaints of fuel impurities and contamination were filed with the company. Customers began demanding refunds and adjustments. This was not a particularly good time for a bad business reputation at XYZ Refinery. Although the company had always prided itself on the quality of its products, the depressed economy had still caused the cancellation of several longstanding accounts. Customers were simply trying to do more with less and XYZ was feeling the results of these cutbacks.

By examining the various complaints, management was able to isolate these problems to one specific area, fuel farm #1. A quick analysis of fuel samples confirmed high levels of metal contamination in the kerosene fuel. On Monday, 15 July 1991, Line Supervisor Ted Whitt assigned three fuel farm technicians to work with an engineer at fuel farm #1. This crew was to examine and inspect the interior of the underground vault. They were to look for any deterioration of the steel tanks inside the vault that could possibly indicate the source of the contamination. Since the vault had not been accessed since its last preventive maintenance inspection in August of the previous year, management was unclear as to the condition of the inside holding tanks.

Jake Boid, fuel farm technician, opened the manhole cover and peered into the tank. He noticed the ladder appeared wet and he could not see to the bottom of the vault. An explosion-proof light was lowered into the vault chamber until the floor was visible. Although there were several puddles of what looked like water, the soil floor of the tank appear to be dry and stable. The faint odor coming from the vault was somewhat musty, but Jake could not detect any scent of fuel. He proceeded to climb down the ladder into the vault, while the other two technicians went to their service truck to retrieve the equipment they would need for the inspection. Engineer Fred Galloway reviewed his inspection procedures one last time. The second technician, Max Wells, climbed down the ladder and immediately noticed that Jake was lying motionless on the ground. He jumped to the floor of the vault and yelled for help. He too fell unconscious to the floor. The third technician, Mark Shaw, already halfway into the manhole at this time, yelled for Fred to get help. Mark slipped off the wet ladder and fell into the vault. He became unconscious almost immediately. Fred peered into the vault and saw two of the three men lying on the floor. Using the two-way radio on the service truck, he called the dispatcher for help. Almost seven minutes passed before rescue personnel began arriving. Because the rescue team had to don self-contained breathing equipment, it was another six minutes before the first of the men in the vault could be retrieved.

The three technicians were pronounced dead at the scene from oxygen deprivation.

Investigation Findings

This tragic accident was the result of a multitude of careless oversights and omissions by management *and* the employees who participated in this event. Because of the complexities of this case, the actual investigation took several months to complete. An accident investigation team, led by a member of executive management, required the services of several safety and health experts, as well as consultants specializing in areas such as microbiology, metallurgy, and hydrology. The following is a listing, in no particular order, of just some of the 76 findings identified during the actual investigation:

1. *Inadequate Procedures:* Fred Galloway's written procedure focused on the inspection of the tanks and did not address the vault entry at all. While a company maintenance procedure did exist that covered the hazards and warnings of confined-space entry, it was not used in this case. In fact, neither the maintenance procedure nor the regular maintenance crew were consulted before writing the inspection procedure.

2. *Management Priorities:* In this case, it was demonstrated that management was more concerned with image and profit than it was with the safety and health of its employees.

There was no visible indication that management ever considered the potential danger of entering a confined space. While it is true that this accident occurred prior to OSHA's Final Rule on Confined Space Entry, OSHA determined in this case that this employer should have recognized that such hazards existed on its premises.

3. *Lack of Proper Equipment:* The technicians assigned to the task were not provided with proper equipment to perform their jobs. Although it was known that the vault had been sealed for 11 months and that there was no source inside the vault for ventilation, the technicians went to the site with no means of ventilating the chamber (i.e., fans). There was no rescue equipment on the truck either.

4. *Training:* It was quite obvious that none of the dead men were properly trained for this task. The engineer was also apparently unaware of the hazards within a confined space.

While many more causal factors were identified, those listed above provide some example of the shortcomings that led to the ultimate loss event. As stated earlier in this appendix, this complex investigation required the assistance of numerous special consultants. Further evaluation of the vault interior by these experts revealed the following additional information:

There was no natural or artificial ventilation or light provided inside the vault. Temperature ranged between 41 degrees F and 59 degrees F.

Excessive moisture was present in the vault; there was moisture on the exterior of the steel tanks and water puddles on the floor surface due to excessive condensation.

Some type of microorganism was present in the vault that was described as having a slime-like, gelatinous texture, yellow-to-red in color, and distributed in various areas (or "pockets") of concentration within the vault chamber.

Oxygen levels in the vault measured between 16 and 19% near the top and 2-5% near the floor surface.

The soil/earthen floor of the vault was discovered to be rich in hydrocarbons, a result of trace fuel leakage from corroded steel tank welds.

From this data, the experts were able to determine that a slime mold identified as *Cladosporium resinae,* more commonly known as the "Kerosene fungus," was able to grow and thrive in the nearly ideal conditions inside the vault. The fuel-rich, low-light, cool, moist, stagnant environment was a perfect breeding ground for the fungus. The rapid development of this fungus contributed to the oxygen deficiency in the vault since the mold (as with most living things) consumes oxygen during the growing process.

The scientific and technical aspects of this case are certainly much more complex than can be described here. However, the purpose of using this case as an example is to demonstrate how a complex investigation can require quite sophisticated analyses in order to identify *all* possible causal factors. The reader is encouraged to review this case study further and attempt to determine corrective actions that should be implemented to preclude the possibility of recurrence. Remember to examine all possible cause factors before concluding the investigation process.

Glossary

The following are common definitions of many of the terms that appear in this text or that may be encountered during the accident investigation and loss control process. When appropriate, the source of an individual definition is referenced parenthetically at the end of the definition.

Accident   An unplanned and therefore unwanted or undesired event that results in physical harm and/or property damage.

Accident analysis   A concerted, organized, methodical, planned process of examination and evaluation of all evidence identified during investigation.

Accident investigation   A methodical effort to collect and interpret facts; a systematic look at the nature and extent of the accident, the risks taken, and loss(es) involved; an inquiry as to how and why the accident event occurred.

Accident potential   A situation comprised of human behaviors and/or physical conditions having a probability of resulting in an accident.

Accident phases   When evaluating the sequence of events that resulted in an accident, the events are divided into the three phases or categories of *pre-contact* (before the accident), *contact* (the accident), and *post-contact* (after the accident). Analysis of the events occurring in each phase facilitates the identification of loss-inducing activities and conditions. Also referred to as the three stages of loss control.

Accident repeater   A person who has been principally involved, regardless of cause, in more than one accident within a predetermined and specified period of time, for example one year.

Accident risk   A measure of vulnerability to loss, damage, or injury caused by a dangerous element or factor (MIL-STD-1574A).

Accident sources   Accidents generally involve one or all of the five elements of *people, equipment, material, procedures,* and the *work* environment, each of which must interact for successful business operations. However, when something unplanned and undesired occurs within either of these elements, there is usually some adverse effect on any one or all of the other elements, which if allowed to continue uncorrected, could lead to an incident or accident and subsequent loss.

Acoustics   The science of sound, including its production, transmission, and effects.

Amelioration   Improvement of conditions immediately after an accident; the immediate treatment of injuries and conditions that endanger people and/or property.

Anthropometric   Relating to human body measurements and modes of action to determine their influence on the safe and efficient operation of equipment.

Antioxidant   A compound added to a substance to reduce deterioration from oxidation; a preservative.

Aperture   A fixed or adjustable opening (diaphragm) in the camera through which light enters. These openings are logarithmic progressions and are calibrated in "*f*" numbers. (Abraben 1994)

Asphyxiation   The depletion of oxygen caused by chemical or physical means. Chemical asphyxiants prevent oxygen transfer from the blood to the body cells. Physical asphyxiants prevent oxygen from reaching the blood.

Audiogram   A graphic or tabular record of hearing level measured at different sound frequencies produced by an audiometer in a control setting.

Basic event   As pertains to fault tree analysis (FTA) and/or the management oversight and risk tree (MORT), a root fault event or the first in the process to have occurred that requires no further development or analysis. Represented graphically as a circle.

Barrier   A control (device, mechanism, structure, sign, etc.) intended to prevent the transfer of energy from one element of a system to another.

Bent   Deformed in a shape deviation from the original line or plane, creased, kinked, or folded.

Binding   Restricted in movement or a tightening or sticking condition resulting from high or low temperatures, foreign materials in the mechanism, surface friction, etc.

Blood priority   Figurative reference to management's approach to accident investigation during the early years. Very simply, if there was no blood spilled, then there was no real priority for any action (or budget), and even less management interest.

Bonding   An electrical conductor, or the act of attaching such conductor, to eliminate a difference in electrical or electrostatic potential that would cause a spark to occur between objects.

Boyle's Law   The volume of a mass of gas is inversely proportional to the pressure, provided the temperature remains the same. Also known as Mariotte's Law.

Boyle's-Charles Law   States that the product of the pressure and volume of a gas is a constant, dependent only upon the temperature.

Brinelled   Defaced or distorted surfaces typically caused by shock of impact between surfaces.

Cardiovascular   The system of the human body, including the heart, vessels, and veins, associated with blood distribution and transmission of cellular nutrients.

Catalyst   A substance that can alter the speed of a chemical reaction without itself being chemically altered by the reaction.

Catastrophic   A loss of extraordinary magnitude in physical harm to people or damage and destruction of property.

Cause   An event, situation, or condition that results, or could result *(potential cause)*, directly or indirectly in an accident or incident.

Central nervous system   The portion of the human control and sensory feedback system consisting of the brain, spinal cord, and tributary nerve endings.

Chafed   Wear damage resulting from friction between two parts rubbed together with limited and usually repeated motion.

Charles Law   The volume of a mass of gas is directly proportional to the absolute temperature, provided the pressure remains the same.

Chain reaction   In chemical or nuclear processes, the energy or by products released cause a continuation of the process; in the analysis of accident cause, the sequence of events that resulted in the accident. See *Domino effect.*

Code of Hammurabi   Prescribed indemnification for major injuries or death in 2100 B.C. Required the punishment for causing the injury or death of another person to match the loss incurred, literally *an eye for an eye.* The Code's primary purpose was to assess blame and provide indemnification (or re-venge), rather than to determine the cause of the accident itself.

Compression   Internal stress created in a material by forces acting inward, in opposite directions, in a manner that decreases the size of the material by closing or tightening its molecular structure.

Conflagration   A fire extending over a considerable area and engulfing a consid-erable amount of property.

Constraint   A restriction affecting the degree of freedom to act or move; a boundary or condition that may dictate performance in other than the desired or intended manner.

Containment   Control of the expansion or propagation of accidental loss. Com-monly used in fire control and hazardous chemical spills.

Contributory event   As pertains to fault tree analysis (FTA) and/or the manage-ment oversight and risk tree (MORT), an event that significantly influences the outcome of the top or primary event. Represented graphically as a rectangle and may also be referred to as a *main event.*

Critical job   A job task within an occupation that has been associated with major loss more frequently than others. It could also be a job where an error has the potential to result in a major loss.

Danger   Term of warning applied to an condition, operation, or situation that has the potential for physical harm to personnel and/or damage to property.

Decibel (dB)   A non-dimensional unit of measure used to express sound levels. It is a logarithmic expression of the ratio between a measured sound and a reference level of sound. The commonly used "A" scale of measurement, expressed as dBA, is a sound level that approximates the sensitivity of the human ear.

Defect   Substandard physical condition, either inherent in the material or created through another action or event.

Depth of field   The distance from the lens of a camera to the farthest subject (infinity) in which the image will appear sharp. "Shallow" depth of field is a result of larger apertures; greater depth of field is achieved when smaller aperture openings are used. (Abraben 1994)

Design load   The weight that can be safety supported by a structure, as specified in its design criteria.

Diagram   A geometric drawing used to explain a fact, a process, the sequence of an activity, or the composition of an element as associated with an accident or incident.

Disability   Any illness or injury that prevents a person from normal human activity, either temporarily or permanently.

Disintegrated   Excessive degree of separation or decomposition into fragments, with complete loss of the original form of the material.

Domino effect   The descriptive term used to illustrate the cause-and-effect relationship one event may have to another. As the term implies, one failure event may result in a sequence of additional failure events unless other forces (such as barriers) are in place to prevent or interfere with this process.

Ear protector   Any device designed to reduce the level of noise passing through a person's auditory system (ear muffs, ear plugs, etc.).

Elasticity   The property or capacity of a material to return to its original shape after being distorted by the application of an external force.

Electric arc   The visible effect of an undesired electrical discharge between two electrical connections; produces burned spots or fused metal.

Expert testimony   The opinion of a person skilled in a particular art, science, or profession, having demonstrated special knowledge through experience and education, beyond that which is normally considered common for that art, science, or profession.

Explosion   A rapid build-up and release of pressure caused by chemical reaction or by an over-pressurization within a confined space leading to a massive rupture of the pressurized container.

Eye protector   A device worn by a person or affixed to equipment to deter harmful substances from contact with the human eye.

Failure Mode and Effect Analysis (FMEA)   An in-depth analysis of possible failures and their resulting effects related to system function and performance *(functional FMEA)* or system hardware and components *(hardware FMEA)*.

Fault Tree Analysis (FTA)   A system safety analysis technique used as an inductive method (top down, from the known to the unknown) to evaluate fault or failure events in a system or process.

Flashpoint   The lowest temperature of a liquid at which there are sufficient vapors given off to form a combustible mixture in the air near the surface of the liquid.

Four P's   In evidence collection, the phrase given to the four common categories of *people, parts, papers,* and *positions.*

General Duty Clause   Refers to Section 5(a)(1) of the Occupational Safety and Health Act of 1970, which states: "Each employer shall furnish to each of his employees employment and a place of employment which are free from recognized hazards that are causing or are likely to cause death or serious physical harm to his employees, and shall comply with the occupational safety and health standards promulgated under this Act."

Guard   A physical device to prevent undesired contact with a source of energy between people, equipment, materials, and the environment.

Hazard   A condition or situation that exists within the working environment capable of causing an unwanted release of energy, resulting in physical harm, property damage, or both.

Human error   The end result of multiple factors that influence human performance in a given situation. An often over-used causal factor finding that, by itself, is not entirely descriptive of true accident cause. Human error is considered more a symptom than a cause. See *Human factor.*

Human factor   Any one of a number of underlying circumstances or conditions that directly or indirectly affect human performance. These include physical as well as psychological factors that can potentially lead a person to make an error in judgment or action (human error) resulting in an accident.

Incident   An occurrence, happening, or energy transfer that results from either positive or negative forces influencing events and that may be classified as an accident, mishap, near-miss, or neither, depending on the level and degree of the negative or positive outcome.

Investigation parameters   As defined by management, the specific considerations that must be evaluated to determine the focus of the accident investigation process. Examples include such aspects as the types of occurrences that will require reporting and investigating, the elements of a business operation or function to be investigated and to what extent, how accidents and incidents shall be formally reported, and what use shall be made of the information reported.

Job Hazard Analysis (JHA)   See *Job Safety Analysis.*

Job Safety Analysis (JSA)   A generalized examination of the tasks associated with the performance of a given job and an evaluation of the individual hazards associated with each step required to properly complete the job. The JSA also considers the adequacy of the controls used to prevent or reduce exposure to those hazards. Usually performed by the responsible supervisor for that job and used primarily to train new employees, the JSA is also an excellent source of *paper evidence* during an accident investigation. Also known as the *Job Hazard Analysis.*

Lessons learned   In the accident investigation process, a formal, documented account or report of both the positive and negative aspects of the operation or task involved in the accident. Intended for use as a tool for the prevention of accident recurrence.

Line and staff organization   In the structure of an organization, those members who are directly accountable and responsible for the daily operations of the enterprise are considered *line* management, with the authority to implement or change company policy and operating procedures. Those who serve as advisors to the line and can only recommend changes are considered *staff* management.

Logic gate   As pertains to the system safety applications of fault tree analysis (FTA) and/or the management oversight and risk tree (MORT), a symbol used to identify the association between events on a logic tree.

Loss   Anything that increases costs or reduces productivity and has any adverse effect on the organization or society resulting from either normal operations or unplanned events (SSDC-11)

Loss control   The overall objective of accident investigation. A management responsibility to prevent or control the occurrence of those events that downgrade performance, negatively impact productivity, or otherwise result in a loss of some nature and degree.

Loss ratio   Term used in the insurance industry. A ratio calculated by dividing the amount of loss(es) by the amount of premium(s). Normally expressed as a percentage of the premiums.

Map   A drawing used to illustrate the physical relationships between the elements of people, equipment, materials, and environmental structures associated with an accident or incident.

Mariotte's Law   See *Boyle's Law*.

Mishap   An occurrence that results in some degree of injury, property damage, or both.

Near-accident   See *Near-miss.*

Near-miss   An occurrence or happening that had the potential to result in some degree of injury, property damage, or both, but did not. Also referred to as a *Near-accident.*

Permissible Exposure Limit (PEL)   An OSHA-mandated value that represents the level of air concentrations of chemical substances to which it is believed that workers may be exposed on a daily basis without suffering adverse effects. PELs are enforceable by law under the Occupational Safety and Health Act of 1970. See also *Threshold Limit Value,* or *TLV.*

Planning process   The foundation of a successful accident investigation and loss control program. A detailed and highly comprehensive effort to prepare for the investigation of all possible or potential accident events that could likely occur during the daily operation of a given business enterprise.

Proximate cause   The cause factor that directly produces the effect without the intervention of any other cause. The cause nearest to the effect in time and space.

Residual risk   That risk that remains after the application or implementation of controls, barriers, or other risk-reducing methods or techniques.

Risk   The likelihood or possibility of hazard consequences in terms of severity and probability of occurrence (Stephenson 1991). The probability of occurrence of a loss-producing event, the chance of loss.

Risk event   An occurrence with the potential to lead to an unwanted event such as an accident or an incident.

Risk management   The process whereby management decisions are made concerning control and minimization of hazards and acceptance of residual risk. This includes the identification, analysis, and evaluation of risk and the selection of the most advantageous method of treating it.

Safety   A measure of the degree of freedom from risk or conditions that can cause death, physical harm, or equipment or property damage (Leveson 1986).

Safety professional   An individual who, by virtue of specialized knowledge, skill, and educational accomplishments, has achieved professional status in the safety field (ASSE).

Severity   In accident analysis, a measure of the degree of loss incurred, such as time away from work, resulting from an accident.

System safety analysis   A detailed, systematic method of evaluating the risk of hazard associated with a given system, product, or program. It utilizes a variety of techniques and approaches to accurately identify, resolve, or control exposure to those hazards.

Tension   Internal stress created in a material by forces acting outward in opposite directions, pulling in a manner to cause extension or stretching of the material, expanding its molecular structure.

Threshold Limit Value (TLV)   Represents the level of air concentrations of chemical substances to which it is believed that workers may be exposed on a daily basis without suffering adverse effects. TLVs are developed by the American Conference of Governmental Industrial Hygienists (ACGIH), which is a non-regulatory agency. TLVs are, therefore, not enforceable by law unless adopted by the authority having jurisdiction. See also *Permissible Exposure Limit*, or PEL.

Top event   As pertains to fault tree analysis (FTA) and/or the management oversight and risk tree (MORT), the primary fault event under analysis. Represented graphically as a rectangle.

Undeveloped event   As pertains to fault tree analysis (FTA) and/or the management oversight and risk tree (MORT), an identified fault event that will not be developed further because its occurrence has been determined insignificant with regard to its effect on the top event, or because insufficient data exists to further evaluate the event, or because the event is too complex for the purpose of a specific evaluation. Represented graphically by a diamond shape.

Unsafe act   Any act or action, either planned or unplanned, that has the potential to result in an undesired outcome or loss (injury, property damage, lost production time, etc.).

Unsafe condition   Any existing or possible condition that, if allowed to continue, could result in an undesired outcome or loss (injury, property damage, lost production time, etc.).

Witness   Any person who has first-hand knowledge of some fact related, directly or indirectly, to an accident or incident.

Bibliography

Abraben, E. 1994. *Point of View: The Art of Architectural Photography.* New York: Van Nostrand Reinhold.

ANSI *Z16.2-1962(R1969). Method of Recording Basic Facts Relating to Nature and Occurrence of Work Injuries.* 1962, revised 1969. New York: American National Standards Institute, Inc.

Bird, Frank E., Jr. 1974. *Management Guide to Loss Control.* Englewood Cliffs, NJ: Prentice Hall.

Briscoe, Glen J. September 1982. *Risk Management Guide* (SSDC-11R1). U.S. Department of Energy, System Safety Development Center: EG&G Idaho, Inc., Idaho Falls.

Considine, Douglas M. 1983. *Van Nostrand's Scientific Encyclopedia, 7th Edition.* New York: Van Nostrand Reinhold.

Eastman Kodak Company. 1989. *Kodak Guide to 35mm Photography.* Rochester, NY: Eastman Kodak Company.

Ferry, Ted S. 1981. *Modern Accident Investigation and Analysis.* New York: John Wiley & Sons.

Johnson, W.G. 1973. *MORT, the Management Oversight and Risk Tree.* Washington, DC: U.S. Atomic Energy Commission.

Knox, N.W., and R.W. Eicher. May 1983. *MORT User's Manual (SSDC-4R2).* U.S. Department of Energy, System Safety Development Center: EG&G Idaho, Inc., Idaho Falls.

Koontz, Harold, and Cyril O'Donnell. 1964. *Principals of Management.* 3rd Edition. New York: McGraw-Hill.

Kuhlman, Raymond L. 1977. *Professional Accident Investigation: Investigative Methods and Techniques.* Loganville, GA: International Loss Control Institute, Inc.

Leveson, N.G. June 1986. Software safety: Why, What, and How? *Computing Surveys* 18(2):125–63.

National Safety Council. 1988. *Accident Prevention Manual for Industrial Operations, 9th Edition.* Chicago: National Safety Council.

Nertney, R.J., and R.L. Horman. 1985. *The Impact of the Human on System Safety Analysis (SSDC-32).* U.S. Department of Energy, System Safety Development Center: EG&G Idaho, Inc., Idaho Falls.

Neufeldt, Victoria, and David B. Guralnik. 1988. *Webster's New World Dictionary, 3rd Edition.* New York: Simon & Schuster, Inc.

Stephenson, Joe. 1991. *System Safety 2000.* New York: Van Nostrand Reinhold.

U.S. Department of Defense. August 1979. MIL-STD-1574A: *System Safety Program For Space and Missile Systems.* Department of the Air Force, U.S. Government Printing Office, Washington, DC.

U.S. Department of Labor. July 1993. *Regulations Relating to Labor-General Industry.* Occupational Safety and Health Administration, Code of Federal Regulations, Title 29, Part 1910. Washington, DC: Office of the Federal Register, National Archives and Records Administration.

U.S. Department of Labor. 1977. *Investigation Accidents in the Work Place: A Manual for Compliance Safety and Health Officers.* Occupational Safety and Health Administration, OSHA Publication Number 2288. Washington, DC: U.S. Government Printing Office, Superintendent of Documents.

Vincoli, J. W. August 1990. Pilot Factor: A Problem in Aviation Safety. *Professional Safety* 35(8):25–29.

Vincoli, Jeffrey W. 1993. *Basic Guide to System Safety.* New York: Van Nostrand Reinhold.

Index

Accident Alert form, 82
Accident investigation
  case studies, 216–223
  checklist for, 212–216
  closing of, 56
  elements of
    analysis of basic causes, 39
    corrective actions, 39
    determination of basic causes,
      37–38
    determination of immediate
      causes, 37
    follow-up of corrective
      actions, 40
    on-site examination of
      witnesses, 36, 37
    reconstruction of accident,
      38–39
    reporting of occurrences to
      management, 35
    study of evidence, 37
  historical view, 6
  legal aspects
    arrival of the scene, 109–110
    attorney-client privilege, 109
    internal legal review, 108–109

    legal proceedings, 113–114
    preservation of evidence, 110
    public relations, 111–113
    report writing, 110–111
  manager, role in, 29–32
  parameters in, 9–10
  parties involved in, 10
  planning of. *See* Planning of
    investigation
  professional organizations
    related to, 197–203
  publications on, 203–206
  safety professional, role in, 32–33
  training seminars related to,
    206–207
Accident investigation kit
  administrative supplies, 134–135
  equipment and devices, 136–138
  items for, 132, 133
  location for, 132
  photographic equipment,
    148–150
  tools, 135–136
Accident investigation report,
  99–107
Accident Report Form, 100–102

appendices, 106
conclusions, 106
findings and recommendations,
    105–106
general information, 100
introduction section, 103
investigation board members,
    listing of, 105
legal aspects, 110–111
management review and
    approval, 107
methodology section, 103, 105
narrative of accident event,
    105
specific information, 100
summary of accident, 105
Accident prevention. *See* Loss
    control program
Accidents
catastrophic accidents, 216
causes
    basic causes, 17–18
    equipment, 14
    immediate causes, 18–20
    materials, 14
    people, 12–14
    procedures, 15
    work environment, 15
definition of, 7
Domino effect, 15–21
and hazards, 7
incidents, 8
losses
    loss control, 21–26
    sources of loss, 11–15
and managerial decisions, 11
minor accidents, 216
near accidents, 8
response to. *See* Response to
    accidents
secondary accidents, 49
severity of, 21

American Society of Safety
    Engineers, 60
Attorney-client privilege, 109

Best's Safety Directory, 60
Blood priority, 6

Chain of custody, evidence, 110
Change analysis, 193
Code of Hammurabi, 6
Contact
    categories of, 20
    minimizing losses, 23–25
    point-of-contact, 33
    stage, 23–25, 71
Control
    business functions related to,
        16–17
    loss control program, 17
Corrective action
    follow-up of, 40
    planning procedures for, 47–48
    post-accident, 39–40

Decision diagrams, 172
Department of Defense, 182
Department of Energy, 182
Domino effect, 15–21, 69–70, 117
    and management control,
        165–166
    phases in
        basic causes, 17–18
        immediate causes, 18–20
        incident, 20
        lack of control, 16–17
        loss of people/property, 21

Educated guesses, 9
Effect, and learning, 85–86
Electrical receptacle tester, 137
Emergencies
    investigation planning for, 53–55
    preparation for, 25, 53–55
Emergency telephone numbers,
        within company, 66–68
Employee assistance program, 125
Employees
    as cause of accidents, 12–14, 18
    records related to, 169
    rehabilitation of, 25–26
Environmental mapping, 177,
        178–179
Environmental Protection Agency,
        50
Equipment
    as cause of accidents, 14
    records related to, 167–168, 169
Errors, 117
    blunders, 118
    human error, 118–120
    meaning of, 18, 118
Evidence
    collection of, 73–75
    fragility of, 73–75, 83
    initial assessment of, 71–72, 72
    legal aspects, 110
    methods in examination of, 37
    photographic evidence, 81,
        159–160
    planning activities, 53, 56
    preservation of, 71, 75–77, 110
Exercise, and learning, 85, 86

Facility, records related to, 167–168
Failure modes, in human
        performance, 124
Failure modes and effects analysis,
        37

Fault tree analysis, 37, 172, 183–190
    construction of, 184–186
    functions of, 183–184
    qualitative and quantitative
        reasoning, 184
    symbols for, 186–190
    top event in, 183, 185
First aid, in rescue operations, 69
Flow diagrams, 171–172
Fragility of evidence, 73–75, 83

Gas and vapor analyzer, 137

Hazards
    elimination of, 24
    modification of situation, 24
    nature of, 7
    positive controls, 24
    reduction of, 24
    separation of persons from, 24
Human error
    and accident prevention, 119
    vs. human factors, 118–119
    and management, 119–120
    steps in process of, 122
Human performance
    basic response model, 121–122
    factors affecting performance,
        121
    failure modes, 124
    quality assurance, 122–123
    stress resistance, 123
    supervisor's role, 124–126

Incidents, nature of, 8
Intensity, and learning, 86

Investigators, 57–63
  number assigned to case, 58
  selection of, 58, 60
  team approach, 58–60
  training of, 60–63

Learning, and memory, 84–86
Legal proceedings, 113–114
Legal requirements, for
    investigation, 53
Loss control program, 17–26
  contact stage, 23–25
    minimization of losses, 23–24
  documentation of lessons
      learned, 107–108
  loss control activities, 70–71
  planning for, 48
  post-contact stage, 25–26
    actions to take, 25–26
  pre-contact stage, 21, 23
Losses, types of, 22

Management
  notification about accident, 50
  policy, records related to, 170
Management Oversight and Risk
    Tree (MORT), 190–195
  analytical chart, 192–193
  change analysis, 193
  event tree, 193–194
  features of, 191
  mini-MORT, 193–194
  uses of, 193
Managers
  decisions and accidents, 11
  identification of evidence, 72
  role in accident investigation,
      29–32

Maps
  environmental mapping, 177,
      178–179
  items for recording, 176
  and measurement, 174–177
  position-mapping, 172, 174, 175
  reconstruction aids, 177, 180
Materials, as cause of accidents, 14
Memory, 84–87
  and laws of learning, 84–86
Miniature camera, 145

National Bureau of Standards,
    60
Near accidents, nature of, 8
Near misses, 23
News media, presentation of
    information to, 79–81
Notification about accident
  emergency telephone numbers,
      66–68
  initial, 66–68
  off-site accidents, 68
  to relatives of injured, 78–79

Occupational Safety and Health
    Administration (OSHA),
    32–33, 50
On-site investigation
  cassette recorder, use at, 136
  and evidence, 53
  legal aspects, 109–110
  planning activities, 51–53
  of witnesses, 36, 37

Papers, as evidence, 75
Parameters, in accident
    investigation, 9–10
Parts, as evidence, 75

Personal protective equipment, 24, 139
PET chart, 194
Photographic equipment
  for accident investigation kit, 148–150
  correct use of camera, 140–141
  miniature camera, 145
  Polaroid camera, 145–146
  35mm single lens reflex camera, 147–148
  video camera, 146–147
Photographic methods
  sources of information about, 150–151
  35mm single lens reflex camera
    automatic cameras, 151
    depth of field, 152–153, 154
    film speed, 153, 155
    lens opening, 152–153
    shutter speed, 151–152
    use of, 151–155
Photographs
  and equipment failure, 159–160
  as evidence, 81, 159–160
  general uses of, 143
  photographic log, 160–162
  planning for, 141–143
  problems in accident photos, 155–159
    camera position, 157
    film color sensitivity, 156
    focal length, 157–158
    light angle effects, 156–157
    shadow effects, 158–159
  as record of injury/damage, 159
  release of, 81
  specific uses of, 144
Planning of investigation
  actions, development of, 45–46
  activities at arrival at accident scene, 51–53

additional activities, 56–57
  for control of emergencies, 53–55
  data analysis phase, 55–56
  discovery activities, 50–51
  establishment of objectives, 43–44
  establishment of priorities, 44–45
  pre-accident activities, 48–50
  preparation of corrective action, 47–48
  schedules, development of, 46–47
Point-of-contact, 33
Polaroid camera, 145–146
Position-mapping, 172, 174, 175
Positions, evidence, 75
Post-traumatic stress disorder, 86
Primacy, and learning, 86
Procedures, as cause of accidents, 15
Professional organizations, related to accident investigations, 197–203
Publications, related to accident investigations, 197–203
Public relations, 77–83
  importance of, 78
  informing relatives of injured, 78–79
  legal aspects, 111–113
  news media, presentation of information to, 79–81
  photographs, release of, 81
  records related to, 170
  release of information within company, 81, 83
Purchasing, records related to, 168–169

Quality assurance, and human performance, 123

Readiness, and learning, 84–85
Recency, and learning, 86
Reconstruction of accident, 38–39
Recordkeeping, types of contact,
    20
Records
    for facility and equipment
        design and maintenance,
        167–168
    for management policy, 166
    for personnel and equipment
        certifications, 169
    for public relations, 170
    for purchasing records, 168–169
    for risk assessment, 166–167
Reenactment, 36
Reporting of accident
    employee participation, 35
    formal policy for, 5
    importance of, 35
    See also Accident investigation
        report
Rescue
    activities in, 69–70
    preparation for, 69
Response to accidents
    accident investigation report,
        99–107
    evidence collection, 73–75
    initial assessment of evidence,
        71–72
    loss control actions, 70–71
    notification actions, 66–68
    preservation of evidence, 75–77
    public relations, 77–83
    rescue actions, 69–70
    witness statements, 83–99
Risk assessment, records related
    to, 166–167
Risk management, importance of,
    8–9
Risks, calculated risks, 9

Safety professional
    pre-accident role, 33
    role in accident investigation,
        32–33, 34
Salvage
    planning for, 56–57
    policy for, 26
Scale models, in reconstruction of
    accident, 38–39
Schedule, of accident investigation,
    46–47
Secondary accidents, 49
    causes of, 49
    prevention of, 70
Sound meter, 136
Stress resistance, in human
    performance, 123
Symbols
    in fault tree analysis, 186–190
    Management Oversight and
        Risk Tree (MORT), 195
Synthetic smoke, 137
System safety applications
    fault tree analysis, 183–190
    Management Oversight and
        Risk Tree (MORT),
        190–195

Team approach, to investigation,
    58–60
Telephone stickers, emergency
    phone numbers, 66–68
35mm single lens reflex camera,
    147–148
    use of, 151–155
Training
    for investigators, 60–63
    seminars in accident
        investigation, 206–207

Unsafe acts
  examples of, 19
  meaning of, 19
Unsafe conditions
  examples of, 19
  meaning of, 19

Video camera, 146–147

Witnesses
  communication with
    investigator, 91
  definition of, 83–84
  first-time statements, 90, 95
  interviews
    areas covered in, 93–94
    beginning interview, 90–93

conclusion of, 94–95
follow-up interviews, 96–99
group interviews, 90
location for, 89–90
note-taking during, 91
post-interview, 95–96
  locating witnesses, 87–88
  memory, 84–87
  on-site examination of, 36
  rationalization of events by,
    96–97
  self-preservation of, 89
  and sensory evidence, 75
  unbiased testimony, 88–89
  Witness Statement Form, 98
  written statement, 97–99
Work environment, as cause of
  accidents, 15, 18
Work-product privileges, 111
Written statements, by witnesses,
  97–99